Unexpected Similarities of the Universe with Atomic and Molecular Systems: What a Beautiful World

Unexpected Similarities of the Universe with Atomic and Molecular Systems: What a Beautiful World

Eugene Oks
Physics Department, Auburn University, Auburn, USA

IOP Publishing, Bristol, UK

ISBN 978-0-7503-1707-8 (ebook)
ISBN 978-0-7503-1705-4 (print)
ISBN 978-0-7503-1706-1 (mobi)

DOI 10.1088/2053-2563/aafed9

Version: 20190401

IOP Expanding Physics
ISSN 2053-2563 (online)
ISSN 2054-7315 (print)

British Library Cataloguing-in-Publication Data: A catalogue record for this book is available from the British Library.

Published by IOP Publishing, wholly owned by The Institute of Physics, London

IOP Publishing, Temple Circus, Temple Way, Bristol, BS1 6HG, UK

US Office: IOP Publishing, Inc., 190 North Independence Mall West, Suite 601, Philadelphia, PA 19106, USA

To Alex, Ellen, and Andrew

Contents

Author biography

Eugene Oks

Eugene Oks received his PhD degree from the Moscow Institute of Physics and Technology, and later the highest degree of Doctor of Sciences from the Institute of General Physics of the Academy of Sciences of the USSR by the decision of the Scientific Council led by the Nobel Prize winner, academician A M Prokhorov. According to the Statute of the Doctor of Sciences degree, this highest degree is awarded only to the most outstanding PhD scientists who founded a new research field of great interest. Oks worked in Moscow (USSR) as the head of a research unit at the Center for Studying Surfaces and Vacuum, then at the Ruhr University in Bochum (Germany) as an invited professor, and for the last 28 years at the Physics Department of Auburn University (USA) in the position of Professor. He has conducted research in five areas: atomic and molecular physics, plasma physics, laser physics, nonlinear dynamics, and astrophysics. He founded/co-founded and developed new research fields, such as intra-Stark spectroscopy (a new class of nonlinear optical phenomena in plasmas), masing without inversion (advanced schemes for generating/amplifying coherent microwave radiation), and quantum chaos (nonlinear dynamics in the microscopic world). He also developed a large number of advanced spectroscopic methods for diagnosing various laboratory and astrophysical plasmas—the methods were then used and continue to be used by many experimental groups around the world. He has published over 400 papers and 5 books, including the books *Breaking Paradigms in Atomic and Molecular Physics* and *Diagnostics of Laboratory and Astrophysical Plasmas Using Spectral Lineshapes of One-, Two-, and Three-Electron Systems*. He is the Editor-in-Chief of the journal *International Review of Atomic and Molecular Physics*. He is a member of the editorial boards of two other journals, *International Journal of Spectroscopy* and *Open Journal of Microphysics*. He is also a member of the International Program Committees of two series of conferences, Spectral Line Shapes as well as the Zvenigorod Conference on Plasma Physics and Controlled Fusion.

IOP Publishing

Unexpected Similarities of the Universe with Atomic and Molecular Systems: What a Beautiful World

Eugene Oks

Chapter 1

Introduction

I wonder if those electrons
Are worlds with oceans and soils,
With all the human passions spectrum
That on the Earth forever boils,
And every atom is a sighting
Of a Universe with planet swarms,
In a small volume full of lighting,
And cosmic hurricanes, and storms.

These are the first eight lines of a poem written in 1922 by the great Russian poet Valery Bryusov—translated by the current author. Below is the Russian original:

Быть может, эти электроны
Миры, где пять материков,
Искусства, знанья, войны, троны
И память сорока веков!
Ещё, быть может, каждый атом –
Вселенная, где сто планет;
Там - всё, что здесь, в объеме сжатом,
Но также то, чего здесь нет.

When the night sky is full of stars, most people (including the author) become fascinated while looking at it. It is so beautiful, it feels romantic. It is the subject of songs and paintings, including the famous Van Gogh painting *The Starry Night*, reproduced in figure 1.1.

doi:10.1088/2053-2563/aafed9ch1

Figure 1.1. Van Gogh's *The Starry Night.*

It is difficult to imagine that at the moment we look at the stars, many of them do not exist anymore, but we still see their light. This is because the speed of light is not infinite and it takes a large number of years (from tens to billions) for the light emitted by a star to reach us, so that the travel time can significantly exceed the lifetime of the star.

This does not diminish the magic of the Universe, but rather enhances both its magic and the emotions it stirs. As the writer Phyllis Curott put it [1]:

> Most people know intuitively that when you fall in love, the world is full of magic. What they don't know is that when you discover the Universe is full of magic, you fall in love with the world.

So, the macro-world—the Universe—is full of magic. What about the micro-world—atoms and molecules? We cannot see them with the naked eye (or even with an ordinary microscope), so for most people they do not cause such emotions as stars. Little do most people know how much alike the micro- and macro-worlds are.

It is amazing that the great Russian poet Valery Bryusov (who was definitely not a physicist) presented the idea of the likeness of the micro- and macro-worlds in his beautiful poem of 1922, chosen as the epigraph to the present book. Bryusov belonged to the Silver Age of Russian poetry, corresponding to the last decade of the 19th century and first two or three decades of the 20th century. It was an extremely creative period in the history of Russian poetry, just as prominent as the Golden Age a century earlier.

Bryusov's poem was one of the primary inspirations for writing the present book. The current author is trying to reveal the hidden beauty of the micro-world by showing many of its mind-boggling similarities and connections to the Universe.

The main part of this book is intended for readers equipped with just curiosity. It does not require even a high school (or lyceum) background in mathematics. More rigorous (and mathematically involved) descriptions of the phenomena discussed in the main part of this book are delegated to appendices, which may interest readers at the level of undergraduate physics, graduate students, and researchers.

The main part of this book is written as a series of essays on the topics listed in the table of contents. The style of the essays was inspired by V V Beletsky's book *Essays on the Motion of Celestial Bodies* published by Birkhäuser/Springer in 2001 (first published in Russian in 1972) [2].

The author presents physical theories explaining the phenomena discussed in this book, and in the main part of the book the theories are presented in a simple way. However, from here on, it is more appropriate to use the term 'theoretical *models*' rather than 'theories'. The reason is that there is not yet any physical 'theory of everything'. Let us explain this in more detail.

To date, there are four known fundamental forces (or interactions) in nature. The two which we encounter most frequently in our everyday life are the electromagnetic and gravitational interactions. Indeed, an overwhelming proportion of modern technology, such as computers, smart phones, tablets and so on (and of course electricity itself), is based on electromagnetic interactions, while the fact that we can walk in the street without the danger of being thrown into outer space is due to our gravitational interaction with the Earth. The two other fundamental forces are the strong force, which makes it possible for atomic nuclei to exist, and the weak force, which manifests in some nuclear processes.

There is a 'theory' that unites three of these fundamental forces: the electromagnetic, strong, and weak interactions. It is more appropriately called the 'Standard *Model*' because it is not the 'theory of everything', since it does not include gravitation.

The Standard Model is also just a model from another point of view. To unite the above three fundamental forces it employs about five dozen (!) 'elementary particles' (including the antiparticles of antimatter). Five dozen is too many—this violates the concept of simplicity, which is the cornerstone of the 'Occam's razor' principle:

> Other things being equal, simpler explanations are generally better than more complex ones.

Throughout its history, Occam's razor has been the guiding principle for theoretical physicists: the simpler hypotheses about nature are more likely to be true. This notion is deeply rooted in the *aesthetics* and *beauty* of physical theories. Here is what the American theoretical physicist, Nobel Prize winner Richard Feynman wrote in his book *The Character of Physical Law* (first published in 1965 by the BBC) [3]:

> You can recognize the truth by its beauty and simplicity. It is always easy when you have made a guess [a physical hypothesis], and done two or three little calculations to make sure that it is not obviously wrong, to know that it is right… The truth always turns out to be simpler than you thought.

The latest experimentally discovered elementary particle (in 2012) is called the Higgs boson, predicted theoretically by the British theoretical physicist and Nobel Prize winner Peter Higgs already in 1964. The Higgs boson is responsible for particles of matter having mass. Incidentally, there is a joke about this particle, as follows:

> A Higgs boson enters a church one Sunday morning and is stopped by the priest. 'We don't allow your kind to enter the Catholic church!' to which the boson replies 'But Father, you can't have mass without me!'

Coming back to physical models, another important thing is their plurality: frequently the same physical phenomenon/experiment can be explained by more than one model. A question might arise as to why keep more than one, say two, theoretical models for the same physical phenomenon. The answer is that when in the future a new experimental result, unexplained by the two models, is produced, it might be relatively easy to adjust only one of the two models to incorporate the new experimental result, while it would be difficult to adjust the other model. (This was also noted by Feynman in his book [3], quoted above.) Thus keeping a plurality of models is important for the future of physics.

As one could probably guess from the preceding discussion, physicists come in two types: theorists and experimentalists. (Nowadays there is also a third kind, computational physicists; the author postpones his comments on this until the final chapter 11 'Concluding remarks'.) Theorists often differ from experimentalists in their philosophy and personality. To develop models that satisfy the Occam's razor principle and can be explained qualitatively 'by the rule of thumb', theorists usually employ *analytical methods*. Thus the outcomes are *analytical models* describing the phenomenon using equations (in distinction to the numerical simulation models of computational physicists—models that are nearly impossible to explain by the rule of thumb because they lack physical insight compared to analytical models). Theoretical physicists typically start from the simplest analytical models, providing only a crude description of the phenomenon, and then proceed to developing more realistic (and therefore more complicated) analytical models. This can be illustrated by the following joke:

> A farmer has problems with his chickens: all of a sudden, they are all becoming very ill. After trying all conventional means, he calls a physicist to see if they can figure out what is wrong. The physicist tries. He stands there and looks at the chickens for a long time without touching them or anything. Then all of the sudden he starts scribbling away in a notebook. Finally, after several gruesome calculations, he exclaims, 'I've got it! But it only works for spherical chickens in a vacuum.'

Here is another joke illustrating the same, but not only the same:

> If a theorist is asked to calculate the stability of an ordinary table with four legs, he would very quickly calculate analytically the stability of a table with one leg and the stability of a table with an infinite number of legs. Then he would spend the rest of his life unsuccessfully trying to calculate analytically the stability of a table with four legs.

There is some truth in the latter joke (as in many jokes). Namely, it is a time-tested idea for theorists to start from so-called 'limiting/asymptotic cases' (such as tables with one leg or an infinite number of legs as in the above joke). It can be relatively easy to develop analytical models for limiting cases (and thus to have a rule of thumb explanation for limiting cases). This can often help to develop a model for the general case of the phenomenon under consideration.

In the course of time, physical experiments become more and more complicated, involving a large number of various types of equipment that are intended to work together, but most of the time do not. Thus, experimentalists can spend more and more of their time trying to make the experiment work, and less and less time actually running the experiment and obtaining results. Here is the corresponding joke:

> *Question*: You enter the laboratory and see an experiment. How will you know which class is it?
> *Answer*: If it's green and wiggles, it's biology.
> If it stinks, it's chemistry.
> If it doesn't work, it's physics.

Here is another one, illustrating the relationship between experimentalists and theorists:

> The experimentalist comes running excitedly into the theorist's office, waving a graph taken off his latest experiment. 'Hmmm', says the theorist, 'That's exactly where you'd expect to see that peak. Here's the reason.' A long logical explanation follows. In the middle of it, the experimentalist says 'Wait a minute', studies the chart for a second, and says, 'Oops, this is upside down.' He fixes it. 'Hmmm', says the theorist, 'you'd expect to see a dip in exactly that position. Here's the reason…'.

The moral: the minds of theorists demonstrate spontaneity, flexibility, and inventiveness.

Coming back to the approximately five dozen elementary particles in the Standard Model—is this not too many to keep in mind? Luckily, for the overwhelming majority of everyday phenomena, it can be sufficient to keep in mind only very few of the most common and least exotic constituents of nature:

light (more rigorously, electromagnetic radiation), represented by particles called photons, and atoms, consisting of negatively charged electrons, positively charged protons, and electrically neutral neutrons (the latter two kinds of particles being inside the atomic nucleus). Atoms are electrically neutral because the total negative charge of the electrons is exactly balanced by the total positive charge of the protons. If one or more electrons are missing, the atom becomes a positive ion. Perhaps the following jokes can help in memorizing these building blocks of Nature:

> A photon checks into a hotel. The bellhop asks: 'Can I help you with your luggage?' It replies: 'I don't have any. I always travel light.'

> Why can't you trust an atom?—Because they make up everything.

> Two atoms rounded a corner from two sides of a building and bumped into each other. One of them said: 'I think I lost an electron.' The other asked: 'Are you sure?' The first one answered: 'I am positive.'

> A neutron walked into a bar and asked, 'How much for a drink?' The bartender replied, 'For you, no charge.'

The author would like to conclude this introduction with few 'philosophical' thoughts. In the past, atoms were considered as indivisible elementary particles of matter. As experiments moved to higher energies, it was found that an atom has a structure: it consists of electrons and the nucleus, both considered as indivisible elementary particles of matter. As experiments moved to even higher energies, it was found that nuclei have a structure: they consist of protons and neutrons, both considered as indivisible elementary particles of matter. As experiments moved further to higher energies, it was found that protons and neutrons have structures: they consist of so-called 'quarks', considered as indivisible elementary particles of matter.

To the author this process resembles the Russian doll (also known as Matryoshka doll), which is a set of wooden dolls of decreasing size placed one inside the other. You open the largest one and find inside it a slightly smaller one. You open it and find inside it an even smaller one, and so on, as pictured in figure 1.2. Some sets of Russian dolls contain over a dozen wooden dolls inside the largest one.

The question is: are quarks (which are the building blocks of the Standard Model) really the smallest 'doll' within the Russian doll of nature or, as experiments proceed to higher and higher energies, will we find smaller and smaller 'elementary' particles of nature? The author is inclined to think that this process does not have a limit, at least as a matter of principle.

Figure 1.2. A Russian doll.

Some more food for thought: as physics has progressed, it has become apparent that we know no more than 10% of what the Universe is made of. We just call the unknowns 'dark matter' and 'dark energy' without any clue what they actually are. Incidentally, it also became apparent that we use no more than 10% of our brain power—an interesting coincidence in numbers. So, would it take being able to engage the full power of the brain to figure out what most of the Universe is made of?

There is an illuminating way to visualize the present situation. Imagine that knowledge occupies a sphere of some radius—the Sphere of Knowledge. In the course of time, as the amount of knowledge increases, the radius and the volume of the sphere increase. But as they do, the surface area of the Sphere of Knowledge also increases, i.e. where it touches the Unknown! In other words, the more we know, the better we realize how much we do not know.

Having written this, the author wishes the readers a happy, fun-filled journey through the rest of this book.

References

[1] Curott P 1999 *Book of Shadows: A Modern Woman's Journey into the Wisdom of Witchcraft and the Magic of the Goddess* (Potter/Ten Speed/Harmony/Rodale)

[2] Beletsky V V 2001 *Essays on the Motion of Celestial Bodies* (Basel: Birkhäuser/Springer)

[3] Feynman R 2017 *The Character of Physical Law* (Cambridge, MA: MIT Press) p 171

Chapter 2

Reasons from astrophysics and atomic physics for our world to be three-dimensional

The space we live in and our Universe are three-dimensional. The objects we encounter in everyday life have three spatial dimensions, such as length, width, and height. Objects that can fly, such as birds or airplanes, can generally move in any direction, but the motion in any direction is just a combination of three fundamental types of motion: forward (or backward), right (or left), and up (or down).

The question why our Universe has three spatial dimensions (rather than two, or four, or n spatial dimensions, where n is any positive integer) has intrigued physicists for dozens and dozens of years. There are numerous approaches to answering the question of why our Universe is three-dimensional, and there are various aspects to this problem. (Here and below we are talking only about spatial dimensions, unless specified to the contrary.)

Let us start from the fact that the stable orbits of planets around stars and the stationary states of electrons in atoms are possible only in a three-dimensional world and in a two-dimensional world. Here is why.

In a three-dimensional world (such as ours), the gravitational force (F) between a star and a planet (or between any two objects) falls off as the inverse square of the distance (r) between them: F is proportional to $1/r^2$. This can be understood by the following reasoning. A massive star creates a gravitational field around it and the field spreads out in an equal way in all directions. If we imagine a sphere of some radius r_1 with the star at the center, the gravitational field of the star acts on each surface element of the sphere. If we then imagine a sphere of some greater radius r_2 ($r_2 > r_1$), the gravitational field of the star acts on each surface element of the greater sphere with a smaller force because the field acts on a greater overall surface of the sphere. The overall surface area (S) of any sphere of radius r is proportional to the square of the radius: S is proportional to r^2 (rigorously, $S = 4\pi r^2$). In proportion to

doi:10.1088/2053-2563/aafed9ch2

the overall surface area, the gravitational field tempers off, so that F is proportional to $1/r^2$.

Similarly, the electric force between a nucleus and an electron (or between any two charges) falls off as the square of the inverse distance between them. In turn, this has to do with the fact that the surface area of any sphere of radius r is proportional to the square of the radius. (In the theory of electric interactions this reasoning is the basis of the so-called Gauss law.)

If we were to imagine a two-dimensional world, the gravitational force between a star and a planet (or between any two objects) would fall off as the first inverse power of the distance between them: F would be proportional to $1/r$. This is because in a two-dimensional world, instead of a sphere, we could imagine a circle of radius r with the star at the center, and the gravitational field of the planet would act equally on each element of length of the circle. Since the perimeter (P) of the circle (also known as the circumference) is linearly proportional to its radius ($P = 2\pi r$), the gravitational field would temper off in the same proportion, so that F would be proportional to $1/r$. Similarly, in a two-dimensional world, the electric force between a nucleus and an electron (or between any two charges) would fall off as the first inverse power of the distance between them.

From the above comparison of a three-dimensional world and a two-dimensional world, one might guess that in a world of N dimensions, the gravitational and electric forces would fall off as $1/r^{N-1}$. Actually this can be rigorously proven, but we skip the proof here.

So what? How does this lead to the fact that the stable orbits of planets around stars and the stationary states of electrons in atoms are possible only in a three-dimensional world and in a two-dimensional world? For the explanation we need to discuss the concept of energy.

Energy is perhaps the most important of the so-called 'conserved quantities'. Conserved quantities are extremely useful tools to figure out the outcome of physical processes without the need to know every detail of the 'machinery' of the processes. In other words, conserved quantities allow us to treat a particular physical process as a 'black box' in the sense that we know what goes into the black box and we can calculate what can come out of the black box, without the need to know what happens inside the black box.

While every conserved quantity has a physical meaning (and we will try to understand this through the example of energy), from the mathematical point of view, to any conserved quantity there is 'attached' a rule as to how to calculate it. So, after we calculate the value of a particular conserved quantity as the input into the black box of a physical process, we would immediately know that the calculation of this conserved quantity at the output of the black box must yield the same value as the input.

Energy is considered to consist of two parts. One part, called 'kinetic energy', is an intrinsic property of a moving object. Kinetic energy (K) conveys how hard the object would hit a target. Kinetic energy is proportional to the mass (m) of the object and to the square of its velocity (v): $K = mv^2/2$.

Another part, called 'potential energy', is the capacity to cause the object to move (and thus to acquire kinetic energy). Potential energy typically depends on the location of the object in a field (in a gravitational field or in an electric field).

Energy is the sum of the kinetic and potential energies. While the energy (E) is conserved in physical processes, its constituents are not conserved. For example, the kinetic energy K of the object can increase at the expense of the decrease of the potential energy (U), but the sum $E = K + U$ of the two parts of the energy remains the same.

It is the concept of potential energy that we need to explain why the stable orbits of planets around stars and the stationary states of electrons in atoms are possible only in a three-dimensional world and in a two-dimensional world. Imagine an ordinary kitchen bowl, such as a soup bowl (figure 2.1).

If you were to place a small ball inside the kitchen bowl next to its inner wall at some height from the bottom and then released the ball, it would move along the inner wall in the vertical plane down to the bottom, and then continue going up along the opposite inner wall almost to the initial height. Then the ball would start going along the inner wall in the same vertical plane down to the bottom, and then continue going up along the initial part of the inner wall almost to the initial point, and then would repeat the cycle. (If there were no friction, then the word 'almost' in the previous sentence could be removed and the cycle would continue an infinite number of times.)

The reason why the ball goes down after being released and acquires some kinetic energy is because at the initial point it had some gravitational potential energy. In this cyclic process (called oscillation) there is a continuous transfer between potential energy and kinetic energy, while the sum of these two kinds of energies would remain almost the same (the sum would remain exactly the same if there were no friction).

So, what does this model have to do with the stability of planets around stars and the stationary states of electrons in atoms in a two- or three-dimensional world? To explain this, we need to introduce one more concept, called 'centrifugal energy'.

Let us modify the experiment with the kitchen bowl. If you were to place a small ball inside the kitchen bowl next to its inner wall at some height from the bottom and then, instead of releasing the ball, gave it a relatively strong push in the horizontal

Figure 2.1. An ordinary kitchen bowl as the model for explaining motion in a gravitational field.

plane, the ball would not go to the bottom: instead it would rotate along the inner wall. If there were no friction, the ball would rotate infinitely and would never reach the bottom—pretty much like a planet would rotate around a star and would not fall into the star (at least within a huge interval of time). Another example is the relative stability of a spinning top toy.

It is the rotation that prevents the ball from falling to the bottom of the bowl, and prevents the planet from falling into the star, and prevents a natural satellite (a moon) or an artificial satellite from falling into its planet. The rotation is a stabilizing factor. The term centrifugal energy mentioned above is related to the kinetic energy of rotational motion. Let us quote a short poem (whose author is unknown to the author of this book):

Gravity is so predictable
We understand just what it does
It makes the meteors fall from space
And keeps the oceans in place.
But what great power stops the Moon
From hitting Earth today at noon?
And what is it that keeps the Sun
From making all the planets one?
They simply play the spinning game,
'Centrifugal' is this force's name.
It keeps the planets in their place
And all the galaxies in space.

There are also jokes about gravity, such as the following:

Gravity is a law. Lawbreakers will be brought down!

The best way to explain the stability (or lack thereof) of planets around a star or electrons around the atomic nucleus is by using a plot of the dependence of the 'effective' potential energy U_{eff} on the distance r from the attractive center (that is, the distance from the star in the case of planets or the distance from the atomic nucleus in the case of electrons). The word 'effective' in this situation means that it is the usual potential energy plus the centrifugal (rotational) energy.

Figure 2.2 presents a plot of the effective potential energy (solid curve) typical for a two- or three-dimensional world (both U_{eff} and r are in arbitrary units). The horizontal dashed line represents one of the possible values of the total energy $E = K + U_{\text{eff}}$. The fact that the chosen value of the total energy is negative ($E = -3$) should not cause any confusion. This is because the effective potential energy U_{eff} in $E = K + U_{\text{eff}}$ is counted from the zero level chosen to be when a particle would be at a very large ('infinite') distance from the attractive center, symbolically denoted as $r = \infty$. Then as the distance r decreases from a very large value, the potential energy also decreases and therefore becomes negative, so that the total energy E could also be negative.

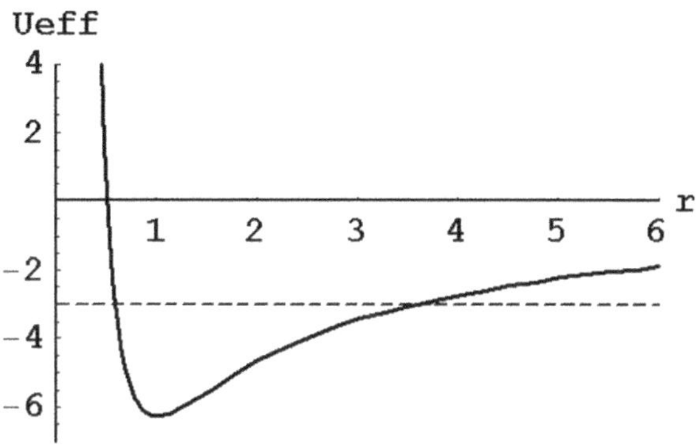

Figure 2.2. Plot of the effective potential energy (solid curve) typical for a two- or three-dimensional world (both U_{eff} and r are in arbitrary units). The horizontal dashed line represents one of the possible values of the total energy $E = K + U_{eff}$.

In terms of the distance r from the attractive center, an object (a planet around a star or an electron around the atomic nucleus) can be imagined as moving along the horizontal dashed line (since the total energy is conserved). The points where the total energy crosses the effective potential energy are called 'turning points'. This is because if the initial distance of the object from the center is within the turning points, then as the object approaches any of the two turning points, it would bounce back toward the other turning point and so on. (The reason is that beyond the turning points, the kinetic energy, being formally expressed as $K = E - U_{eff}$, would be negative, while by the definition of the kinetic energy $K = mv^2/2$ it cannot be negative.) In other words, in terms of the distance r from the attractive center, the object would be confined between the two turning points. In particular, the object cannot reach $r = 0$, that is, it cannot fall into the attractive center. Also the object cannot reach infinitely large distances from the center. This is why in a two- or three-dimensional world, planetary systems and atomic systems can be stable.

In contrast to this, in worlds with more dimensions than three, a typical plot of the effective potential energy looks like the solid curve in figure 2.3. The horizontal dashed line represents again one of the possible values of the total energy $E = K + U_{eff}$.

It is seen that in the physically allowed region, that is in the region of non-negative kinetic energy—the region to the left of the turning point—the object can reach $r = 0$, that is, it can fall into the attractive center. This is why planetary systems and atomic systems can be stable *only* in a two- or three-dimensional world.

Now we are still faced with the question of why our world is three-dimensional rather than two-dimensional. There are several schools of thought on addressing this question.

The argument by one school of thought is that life is possible in the three-dimensional world, but is impossible in the two-dimensional world. Living things (creatures) must have some kind of 'digestive system'. More accurately, they must use energy, so they must take some energy in ('eating') and expel the waste out. In other words, the creatures must have a 'digestive tract'. In a two-dimensional world such a digestive channel would cut the creature into two parts—see figure 2.4.

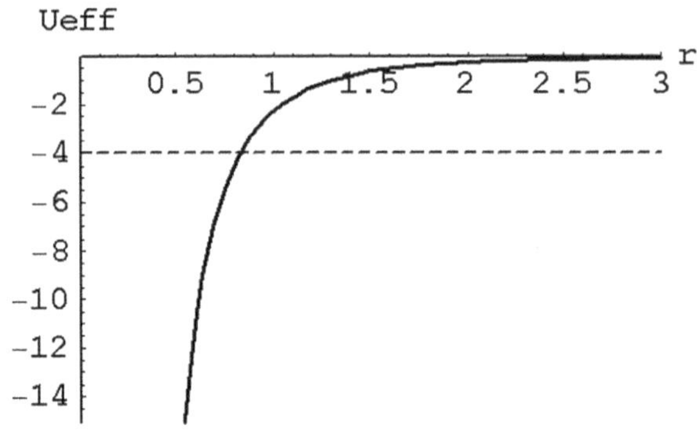

Figure 2.3. Plot of the effective potential energy (solid curve) typical for a world with more dimensions than three (both U_{eff} and r are in arbitrary units). The horizontal dashed line represents one of the possible values of the total energy $E = K + U_{\mathrm{eff}}$.

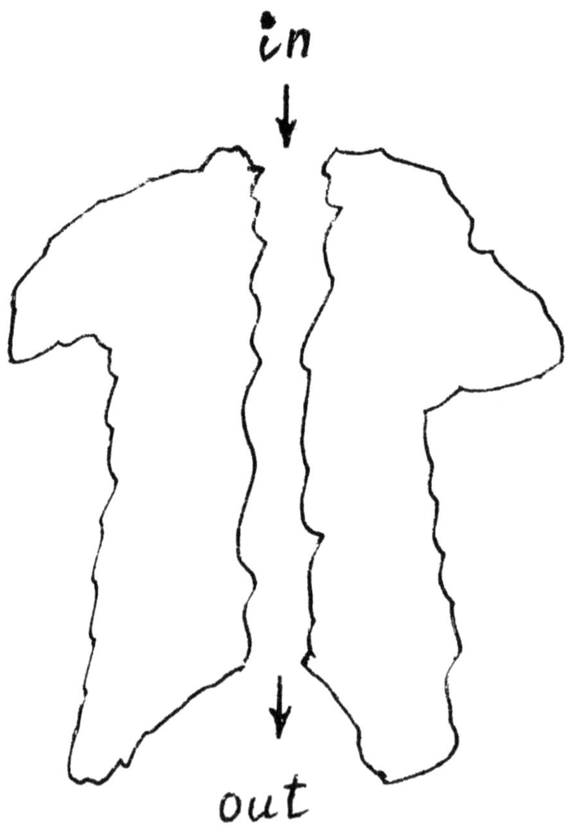

Figure 2.4. A two-dimensional creature with a 'digestive channel' cutting it into two parts.

In order to not cut the creature into two parts, it is necessary for each small part of the digestive channel to be a cylinder wrapped around the substance inside the channel, and cylindrical forms require the third dimension; the same is true also for circulating some kind of 'blood' within the body.

The life-based argument has deficiencies. For example, two-dimensional creatures could live in pairs, like the two halves in figure 2.4. While receiving 'food', they could join each other at the end of the 'digestive' channel. Then, to expell the waste, they could join each other at the beginning of the digestive channel and disjoin at the end of the digestive channel.

There is also a less exotic option. A creature could take the 'food' in and expel the waste out through one and the same opening—regardless of how unpleasant this thought is from the aesthetic point of view. In fact, even in our three-dimensional world there are creatures who do this. The most common example is the jellyfish.

Therefore, it is possible to imagine a two-dimensional 'jellyfish' living in a two-dimensional world. Figure 2.5 depicts a possible family of two-dimensional 'jellyfish' living in a two-dimensional 'house'.

Figure 2.5. A family of the two-dimensional 'jellyfish' living in a two-dimensional 'house'.

It is interesting to note the following. Since in figure 2.5 the 'jellyfish' family is surrounded by the walls of their two-dimensional house, they think that they cannot be seen from the outside. However, we, living in the three-dimensional world, could easily see them from our additional third dimension—their two-dimensional walls are not an obstacle for us. Similarly, when we are in a three-dimensional house, surrounded by the walls, the floor, and the ceiling, we think that we cannot be seen from the outside. However, from the additional fourth dimension (if it did exist), we could easily be seen: our three-dimensional barriers (the walls, the floor, and the ceiling) would not be an obstacle for a four-dimensional or a higher dimensional creature.

Among the life-based arguments for excluding the two-dimensional world, there is also one based on Einstein's theory of gravitation (the theory frequently called general relativity). The gist of this theory is that a massive body (such as a star) does not exert gravitational force on a light body (such as a planet). Rather, the massive body curves the space around it, making the space like the kitchen bowl in figure 1.1 with the massive body at the bottom of the bowl. Then a light body placed on the inner surface of the bowl close to the rim would either fall to the massive body or rotate around it if given a sufficient push in the horizontal plane. In other words, what we would perceive as the attraction of the two bodies is just the result of the curvature of space caused by the massive body.

A humorous illustration of Einstein's theory of gravitation was provided by Bruce Elliot in the following short poem entitled 'Brief History of Gravity':

It filled Galileo with mirth
To watch his two rocks fall to Earth,
He gladly proclaimed:
'Their rates are the same,
And quite independent of girth!'
Then Newton announced in due course
His own law of gravity's force:
'It goes, I declare,
As the inverted square
Of the distance from the object to source.'
But remarkably, Einstein's equation
Succeeds to describe gravitation
As space-time that's curved
And it's this that will serve
As the planet's unique motivation.

There are also jokes about general relativity, such as the following:

Question: How many general relativists does it take to change a light bulb?
Answer: Two. One holds the bulb, while the other rotates the Universe.

It could be noted that Einstein developed general relativity several years after he developed special relativity, commonly known simply as the theory of relativity.

The primary outcome of special relativity is that for objects moving with velocities comparable to (though still smaller than) the speed of light, the spatial dimensions and the pace of time change (the latter effect is called 'time dilation'). In fact, the theory of relativity could have been called the theory of absoluteness. Indeed, it postulates that the speed of light is the absolute concept—it is the same for any two observers regardless of the speed with which they move with respect to one another. All the results of the special theory of relativity are consequences of this postulate.

Of course, there are also jokes about special relativity, such as the following:

> A student recognizes Einstein on the train and asks: 'Excuse me, professor, but does New York stop by this train?'

Coming back to general relativity—how does it exclude a two-dimensional world within the life-based approach? It turns out that according to general relativity, there would be no gravitation in a two-dimensional world. Therefore, two-dimensional 'planets' would not orbit two-dimensional 'stars'. Without a stable orbit, the overwhelming majority of planets would not receive a relatively steady stream of energy from the star, required for the existence of life. However, as an exception, a planet could pass by a star (along a rectilinear path) very slowly. So, within an interval of time, during which the distance between the planet and the star varies just a little (say, by about 10%), two-dimensional creatures (such as the two-dimensional jellyfish) could come to life and then become extinct after that interval of time. Such bursts of life might be possible because electron orbits around atomic nuclei would be still stable in a two-dimensional world—in distinction to the orbits of planets around stars.

The most important deficiency of the life-based arguments is that a universe does not have to be 'designed' to support life. Thus, these arguments cannot exclude the possibility of a lifeless two-dimensional world.

There are other schools of thought that provide the most convincing arguments for excluding the possibility of a two-dimensional world. One of them is based on so-called string theory. The motivation behind string theory (and its generalization called brane theory, briefly discussed below) was to design a 'theory of everything', that is to provide a unified description of gravitation and of the other three fundamental forces discussed in chapter 1 of this book. The basic idea of string theory is that the point-like objects of particle physics are just cross-sections of one-dimensional vibrating strings by some kind of surface. As the strings move, so do their cross-sections on this surface, which we perceive as moving point-like particles.

As for the theory of branes, it generalizes the concept of point-like particles to higher dimensions. The string is then a brane of dimension one. A two-dimensional brane is a vibrating surface (the word 'brane' originates from the word 'membrane', which is related to a two-dimensional brane)—and so on to higher dimensional branes.

To create the 'theory of everything', this model requires at least nine spatial dimensions. Further, this model states that currently we see only three dimensions because the other six dimensions are extremely tiny. To visualize the latter concept, let us consider the example of two 'big' dimensions and one 'small' dimension. Let us consider a disk with a radius of several kilometers, but with a thickness of less than 1 mm. All perceived movements of any objects in this (2+1)-world would be along the surface of the disk and no perceived movements in the direction perpendicular to its surface. Thus, the string/brane theory claims that everything in our Universe (including ourselves, which is mind-boggling!) are at least nine-dimensional objects, but we do not perceive six out of these nine dimensions because they could be as thin as, for example 1/1 000 000 000 000 (one trillionth) part of centimeter.

The following verses from the parody song *N-dimensional Space* by Phil Alexander could be in order:

> Like an angle in a triangle,
> A plane upon a plane,
> Euclidean geometry's
> Not hard to explain.
> Like a snowball down a mountain
> Or a path upon a sphere,
> Once you hit that third dimension
> It all becomes less clear.
> Like a tiny 1D superstring
> That is infinitely small,
> It exists in ten dimensions
> Or is it really there at all?
> There are problems that you face
> When pondering *n*-dimensional space.

The ten dimensions mentioned in this song consist of the nine spatial dimensions discussed above plus one time dimension. Altogether it is called ten-dimensional hyperspace. There are jokes about it, such as the following:

> I just got off a plane where I sat next to a string theorist. I asked him how he was able to visualize things in ten-dimensional hyperspace. He replied: 'That's easy. First I visualize it in *n*-dimensional hyperspace and then I just let *n* equal 10.'

So, what does string/brane theory have to do with the proof that two-dimensional worlds cannot exist? In 2005, physicists Karch and Randall published an illuminating paper in the journal *Physical Review Letters* [4]. They considered a nine-dimensional Universe containing branes of all possible dimensionalities: from one-dimensional branes (which are strings) to nine-dimensional branes. (Here an *n*-dimensional brane means a world with *n large* dimensions.) Karch and Randall demonstrated that in the course of the evolution of such a universe, branes of seven out of nine dimensionalities collided and destroyed each other. These authors showed that the 'survivors' are only three-dimensional branes (surprise!) and seven-dimensional branes. However, in seven-dimensional branes, electrons would not have stable orbits around atomic nuclei (as we discussed previously), so stable seven-dimensional branes cannot exist. Thus, the only stable survivors are the three-dimensional branes and our world is one of these three-dimensional branes.

Could there be an argument that there exists more than one three-dimensional brane? The answer is yes. Namely, currently the only way to explain why the gravitational force between two electric charges is an astronomically large number of times weaker than the electric force is the following. There are several surviving three-dimensional branes (at least two) *in the higher dimensional* Universe. The carriers of the three fundamental forces (other than gravitation) are confined within the brane. However, the carrier of the gravitational force (called the graviton) can travel between the branes *through the higher dimensions*. The gravitons spend most of their time traveling between the branes, while spending only a miniscule fraction of time within the brane. Therefore, the strength of the gravitational force perceived within the brane is extremely small compared to the strength of the other three fundamental interactions, which are confined within the brane. However, recently, a large group of scientists (over 1000 people) observed both gravitational and electromagnetic waves from a neutron star collision, and they found that such a 'gravitational leakage' into higher dimensions was not happening [1].

The perceived weakness of the gravitational force is mentioned by Bruce Elliot in the following limerick:

The graviton's something unique,
A particle many would seek,
To Earth we are stuck,
It seems, more by luck,
For its coupling's exceedingly weak!

It should be clarified that 'coupling' refers to the interaction between two objects.

There is yet another school of thought providing a thermodynamics-based argument for our world to be three-dimensional. Thermodynamics is a part of physics that does not deal with individual atoms and molecules, but rather with macroscopic properties of materials (consisting of a very large number of atoms or molecules) that can be observed in everyday life. In particular, thermodynamics specifies the direction in which the macroscopic system can evolve. In thermodynamics there is also a concept of the so-called Helmholz free energy, which provides a measure of the maximum amount of work that the system can perform at constant temperature and volume. It is important to emphasize that for a system at constant temperature and volume, the Helmholtz energy is *minimized* at equilibrium.

To illustrate the thermodynamic point of view, here is a story that is purported to be true:

Nernst, the author of one of the laws of thermodynamics, was breeding fish in a pond near his cottage.

'Why do you bother with them?' asked his acquaintance. 'Even poultry breeding seems to be more interesting?'

> 'I breed animals which are in thermodynamic equilibrium with the environment. Breeding poultry just means warming the Universe at your expense', replied Nernst.

In 2016, physicists Gonzalez-Ayala, Cordero, and Angulo-Brown published a very interesting paper in the journal *European Physics Letters* [2]. They considered an early Universe with an arbitrary number of dimensions. They showed that the *minimum* of the Helmholz free energy per unit volume occurs for the three-dimensional Universe (surprise again!). In other words, out of all possible dimensions, the three-dimensional Universe turned out to be the most favorable on the basis of (roughly speaking) energy considerations.

The above paper [2] dealt with an early Universe, meaning very shortly after the so-called Big Bang that gave birth to the Universe. Incidentally, what caused the Big Bang? Here is a joke:

> A physics student asked his professor what caused the Big Bang. The professor replied: 'The concentration of nothing was so high that it exploded like crazy.'

Thus, all jokes aside, all of the above explains why our world is three-dimensional. But we could still entertain ourselves by considering the amazing features of hypothetical (although not probable) higher dimensional worlds. Let us start from our three-dimensional world and consider a stool, like the one depicted in figure 2.6. The minimum number of legs for a stool in the three-dimensional world is three. Indeed, if it were to have two legs, a person trying to sit on it would not move in the (two-dimensional) plane containing the two legs, but would fall in the direction perpendicular to that plane, that is, would fall in the third dimension.

In a four-dimensional world, a person trying to sit on a stool with three legs would fall in the direction of the fourth dimension. (This is similar to the two-legged stool in the three-dimensional world: it would fall in the direction of the third dimension). So, in a four-dimensional world, a stool must have at least four legs, in a five-dimensional world at least five legs, and so on: in an n-dimensional world a stool must have at least n legs. If a 'human being' in the n-dimensional world were to still

Figure 2.6. A stool.

have only two feet (as we do), then the base of each foot should have $(n - 1)$ dimensions for stability.

By the way, in our three-dimensional world, a stool or table with three legs does not rock even on an uneven surface—in contrast to the case of four legs. This is because one of the axioms of Euclidean geometry states that on any three points not located in the same straight line, it is possible to span (to 'pull on') one and only one plane. This is not true for four points, which is why a stool or table with four legs would rock on an uneven surface.

Another example of the interesting features of not three-dimensional worlds concerns the propagation of waves. Waves in a two-dimensional world can be visualized as the waves on the surface of water. If you drop a ball in the water at some point, you would see circular concentric waves propagating away from that point. Far from the source of the disturbance, the waves would seem almost parallel to each other, as in figure 2.7.

If in a two-dimensional world we were to consider the propagation of sound from some source that released a single short pulse of sound, it would be similar to the propagation of waves on the surface of water. Namely, instead of just a single pulse, we would hear a sequence of pulses like waves on water. If the source were to release a continuous melody (such as a song), then we would hear more or less distinctly only the first line of the song. The first wave of the second line of the song would be superimposed with the second wave of the first line of the song; the first wave of the third line of the song would be superimposed with the second wave of the second line of the song and with the third wave of the first line of the song, and so on. The song would become incomprehensible for the listener. The same would be true for speaking: two-dimensional 'human beings' would not be able to comprehend what other two-dimensional 'human beings' said.

It turns out that the same situation would be the case in any world with an *even* number of dimensions: in two, four, six dimensions, and so on. But in worlds with an *odd* number of dimensions (three, five, seven dimensions, and so on), there would be no such problem. So, for example, in a seven-dimensional world (which is one of the

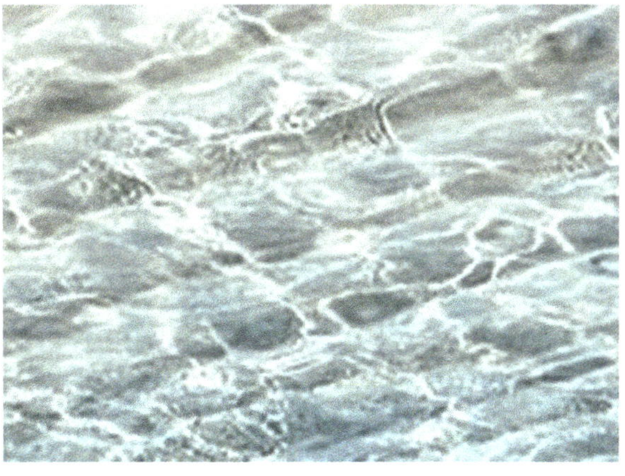

Figure 2.7. Sea waves.

two types of survivors of the collisions of multi-dimensional branes, the other surviving type being a three-dimensional world, as discussed above), the sound would propagate pretty much as in our three-dimensional world. It should be recalled, however, that a seven-dimensional world could exist only for a fleeting moment because electrons would not have stable orbits around atomic nuclei in such a world.

Finally, let us discuss an interesting exercise in multi-dimensionality by Michael Blaber, Professor of Biomedical Sciences [3]. If a circle (which is a two-dimensional object) were to pass through a one-dimensional world, then a one-dimensional observer in that world would see first that a dot appears, then it splits in two dots that move apart, then they come back toward each other, merge in a single dot, and the dot disappears—see figure 2.8, where the observer is indicated by a short blue segment.

If a spherical shell (which is a three-dimensional object) were to pass through a two-dimensional world, then a two-dimensional observer in that world would see first that a dot appears, then it becomes a circle of increasing radius, then the circle starts shrinking, becomes a dot, and disappears—see figure 2.9, where the observer is depicted in blue.

By analogy, it is easy to figure out that if a four-dimensional spherical shell passes through our three-dimensional world, then we would see first that a dot appears, then it becomes a three-dimensional sphere of an increasing radius, then the sphere starts shrinking, becomes a dot, and disappears—see figure 2.10, where the observer is depicted in blue.

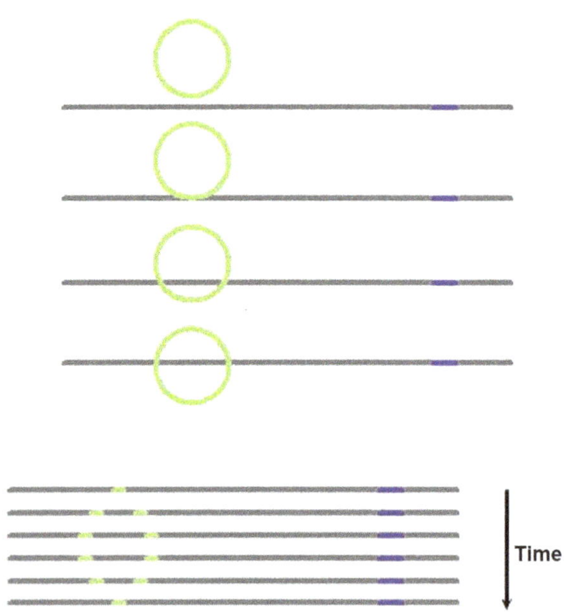

Figure 2.8. A circle passing through a one-dimensional world. (Reprinted with permission from [3].)

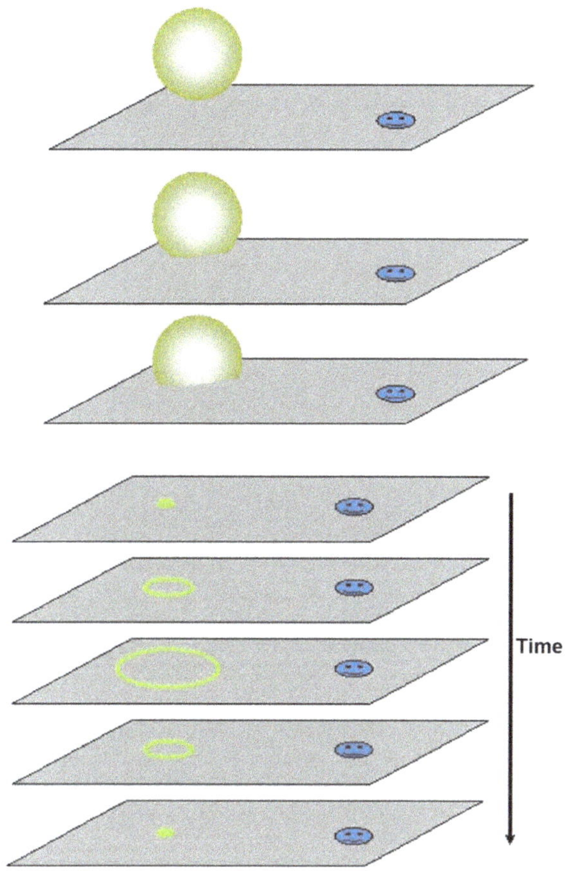

Figure 2.9. A spherical shell passing through a two-dimensional world. (Reprinted with permission from [3].)

Thus, from this exercise it seems possible that what we perceive as the time evolution of an object could be just a passage of a four-dimensional non-evolving object through our three-dimensional world. In particular, from this point of view, it seems possible that the aging (time evolution) of a person that we observe could be just a passage of a four-dimensional non-aging entity (creature?) through our three-dimensional world.

So, if a n-dimensional spherical shell passes through an $(n - 1)$-dimensional world several times, then each time an observer in that world would see the *same* time evolution of the corresponding $(n - 1)$-dimensional sphere. What if an n-dimensional spherical shell passes through an $(n - 2)$-dimensional world? Let us start from a three-dimensional spherical shell passing several times through a one-dimensional world. Then, as noted by Michael Blaber, a one-dimensional observer could see *different* time evolutions of the corresponding one-dimensional object each time. Indeed, if the spherical shell passes through the one-dimensional world in such a way that the center of the shell passes through the one-dimensional world, then the observer would see the two dots moving apart a relatively large distance (equal to

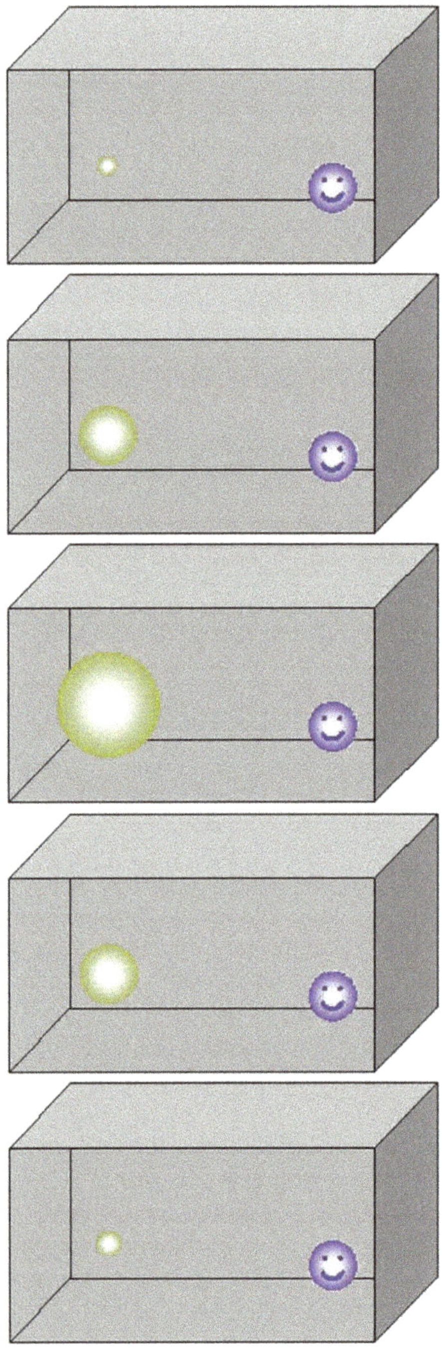

Figure 2.10. A four-dimensional spherical shell passing through our three-dimensional world. (Reprinted with permission from [3].)

the diameter of the shell) before they would come back toward each other and disappear. However, if the spherical shell passes in such a way that the one-dimensional world crosses only a small part of the shell (just a little more than simply touching the shell), then the observer would see the two dots moving just a little bit apart before coming back toward each other and disappearing. Also in this scenario the entire phenomenon of moving dots would last less time than in the previous scenario.

Then by analogy, if a five-dimensional non-evolving object passes through our three-dimensional world several times, we would see *different* evolutions of the corresponding three-dimensional object each time. From this point of view, it seems possible that not only could the aging (time evolution) of an individual person, but also the entire historical evolution of humans from *Australopithecus* to *Homo sapiens* ('wise man' in Latin) be speculated as the passage of a *single* five-*dimensional non-evolving* (non-aging) entity (creature?) through our three-dimensional world. Moreover, the same might relate to all life.

Of course, the major flaw in this reasoning is the neglect of the fact that in four and five dimensions, electrons cannot have stable orbits around atomic nuclei. Consequently, such four- or five-dimensional objects cannot be stable and thus cannot be non-evolving—they could exist only for a fleeting moment.

References

[1] LIGO Scientific Collaboration and Virgo Collaboration 2018 *Tests of General Relativity with GW170817* Report No. LIGO-P1800059 arXiv: 1811.00364

[2] Gonzalez-Ayala J, Cordero R and Angulo-Brown F 2016 *Eur. Phys. Lett.* **113** 40006

[3] Blaber M 2006 The Possibility that All Life is a Single 5th-Dimensional Entity http://mikeblaber.org/dimensions/dimensions.htm

[4] Karch A and Randall L 2005 *Phys. Rev. Lett.* **95** 161601

Chapter 3

More on the stability of atoms and planetary systems in our three-dimensional world: why do we live longer than a tiny fraction of a second?

In chapter 2 the stability of electron orbits around nuclei and planetary orbits around stars was explained with reference to the plot of the effective potential energy $U_{\text{eff}}(r)$ in figure 2.2. The plot of $U_{\text{eff}}(r)$ has a minimum and this allows an object (an electron or a planet) to be 'trapped' between the two turning points, which is why such a shape of $U_{\text{eff}}(r)$ is called a 'potential well'.

Now we will discuss subtleties that could cast a shadow of doubt over the above claim of stability and lead to a more rigorous concept of *relative stability*. First of all, the above reasoning is valid only for an *isolated system*: just one electron orbiting a nucleus or just one planet orbiting a star. Clearly, in reality nothing in the Universe is an isolated system. For the case of one electron orbiting a nucleus there are other atoms and/or electrons nearby. For the case of a planet orbiting a star, there are other stars and/or planets nearby. So, the concept of an isolated system is just an approximate model, which is useful only if the effect of other systems on the system under consideration is relatively small (because other systems are sufficiently far away). Under the action of an external system, the system under consideration could become unstable. For example, an atom could lose an electron and become a so-called positive ion; a planet could be thrown out of its 'solar' system.

However, the most important thing is that, even when disregarding the effect of external systems, the system under consideration could have a finite lifetime, while the word 'stability' implies an infinite lifetime. Here are the details.

First, any electric charge moving with acceleration radiates electromagnetic waves and therefore loses energy. Second, any object orbiting an attractive center moves with acceleration—even if it moves with a constant by magnitude velocity in

doi:10.1088/2053-2563/aafed9ch3

a circular orbit, there is still an acceleration (the centripetal acceleration in this case). Therefore, an electron orbiting an atomic nucleus radiates electromagnetic waves, loses energy, and eventually falls to the nucleus; so, the atom has a finite lifetime—it is unstable. For example, according to classical physics, for a hydrogen atom (one electron around one proton), the corresponding lifetime is only about 10^{-11} s, that is 1/100 000 000 000 of a second (ten trillionths of a second). Thus, classical physics fails to explain the stability of atoms, molecules, everything around us, and (most importantly) ourselves—we remain stable much, much longer than 10^{-11} s. (Even when person A calls person B 'unstable' and recommends going to a psychiatrist, person A does not mean that person B would decay in 10^{-11} s.)

Similarly (but not identically), a planet orbiting a star radiates gravitational waves, loses energy, and eventually would fall to the star. However, there are very significant differences between this case and the case of an electron orbiting a nucleus, as follows.

First, an electromagnetic wave is the propagation of electric and magnetic fields oscillating in the directions perpendicular to the direction of propagation (that is, varying in magnitude in a regular manner with respect to the direction of propagation). In contrast, a gravitational wave is the propagation of ripples in the geometry of space (and of time); when a gravitational wave reaches an object on Earth, the dimensions of this object oscillate very slightly during the passage of the wave.

Second, electromagnetic radiation is an astronomically large number of times more powerful than gravitational radiation. This is because of the following. The power of the electromagnetic radiation P_{EM} is proportional to the square of the electric charge q, to the square of its acceleration a, and inversely proportional to the cube of the speed of light $c = 3 \times 10^{10}$ centimeters (cm) per second (c = 300 000 km s^{-1}, where 'km' stands for kilometer and 's' for second):

$$P_{EM} \backsim q^2 a^2 / c^3. \tag{3.1}$$

Here and below the symbol $\backsim$ stands for 'of the order of'. This means that the corresponding physical quantity (P_{EM} in this case) is estimated by the order of magnitude, within the accuracy of some coefficient on the right-hand side of the equation, and this coefficient is in the range between 1/3 and 3.

Let us recall that acceleration is the rate of change of velocity and velocity is the rate of change of the coordinate. In mathematics, the rate of change of any physical quantity per unit time is called the time derivative—velocity is the time derivative of the coordinate ($v = \mathrm{d}r/\mathrm{d}t$) and acceleration is the time derivative of the velocity ($a = \mathrm{d}v/\mathrm{d}t$), so acceleration is the second time derivative of the coordinate ($a = \mathrm{d}(\mathrm{d}r/\mathrm{d}t)/\mathrm{d}t$ denoted as $\mathrm{d}^2r/\mathrm{d}t^2$). The meaning of these notations can be understood as follows: for example, $v = \mathrm{d}r/\mathrm{d}t$ means that the velocity is the ratio of the change of the coordinate of the object during a very small time interval $\mathrm{d}t$.

The product of the coordinate r of the electron (calculated with respect to the nucleus) and the electron charge e is called the electric dipole moment $d = er$. So, the power of electromagnetic radiation is controlled by the rate of change of the rate of change of the electric dipole moment, that is, by the second time derivative of the electric dipole moment. A non-zero electric dipole moment is possible only because there are two types of electric charges (positive, like protons, and negative, like electrons), so that the electric force, being proportional to the product of the charges of the two interacting objects, could be attractive (between unlike charges) or repulsive (between like charges).

In contrast, the gravitational force is only attractive because it is proportional to the product of masses of the two interacting objects and there are not two kinds of masses (positive and negative), but only one kind, positive. (This is true for the matter known to us so far, which is less than 10% of the matter in the Universe; there is evidence of anti-gravitation in the Universe, which is commonly called 'dark energy' and this is not understood yet.) For this reason, there is no gravitational analogue of the dipole moment. Systems of gravitating masses can have only a 'higher-order' moment called the quadrupole moment Q. For example, for a planet orbiting a star, the quadrupole moment is a quadratic combination of the three mutually perpendicular x-, y-, and z-coordinates of the planet multiplied by the mass, M, of the planet.

The dipole moment is a measure of the distribution of charges or masses along some axis, while the quadrupole moment is a measure of the distribution of charges or masses in a plane. The field produced by a quadrupole configuration falls off with the distance from the system more rapidly than the field produced by the dipole configuration.

The power of the gravitational radiation P_G is proportional to the square of the third time derivative of the gravitational quadrupole moment Q (that is, to the square of the rate of change of the rate of change of the rate of change of Q) and inversely proportional to the fifth power of the speed of light c:

$$P_G \backsim G(d^3Q/dt^3)^2/c^5. \tag{3.2}$$

Here $G = 6.674 \times 10^{-8}$ cm^3 g^{-1} s^2 is the gravitational constant (one of the fundamental physical constants) and 'g' stands for gram (the unit of mass). The factor d^3Q/dt^3 (the third time derivative of Q) in the above equation means that the power of the gravitational radiation depends not only on the acceleration of the masses in different directions, but also on the so-called 'jerk', which is the rate of change of the acceleration.

The comparison of the equation (3.2) for P_G with the equation (3.1) for P_{EM} shows the reasons why a planet around a star loses energy through gravitational radiation much more slowly than an electron orbiting a proton loses energy through the electromagnetic radiation. First, the equation for P_G contains in the denominator the fifth power of the speed of light c while the equation for P_{EM} contains in the denominator only the third power of c. Since the speed of light is numerically very large ($c = 3 \times 10^{10}$ cm), then this factor alone makes the ratio P_G/P_{EM} extremely small. Second, the equation for P_G contains the gravitational constant G, which is numerically small (of the order of 10^{-7} cm^3 g^{-1} s^2).

Third, each time derivative adds to the denominator a factor of the order of the period of revolution $T_{electron}$ of the electron around the proton or of the period of revolution T_{planet} of the planet around the star. So, the equation (3.1) for

P_{EM} contains in the denominator $(T_{electron})^4$, while the equation (3.2) for P_G contains in the denominator $(T_{planet})^6$. Since $T_{electron}$ is of the order of 10^{-16} s, while T_{planet} is of the order of a year (that is T_{planet} is of the order of 10^7 s), then the ratio P_G/P_{EM} contains also the factor $(T_{electron})^4/(T_{planet})^6$, which is of the order of 10^{-106}.

Thus, even though the quadrupole moment Q in the equation for P_G contains the mass M of the planet, which is a relatively large quantity (for example, the Earth's mass is about 6×10^{27} g), so that P_G is proportional to M^2, this large factor cannot overcome the product of the three very small factors described above. Therefore, while according to classical physics, the lifetime of hydrogen atoms should be just about 10^{-11} s due to the loss of energy through electromagnetic radiation, the effect of gravitational radiation on shortening the lifetime of a planet around a star is negligibly small compared to other factors limiting the lifetime of a planet around a star (such as, for example, the lifetime of the star itself—stars gradually 'burn out').

Astrophysicists succeeded in detecting a gravitational wave for the first time only in 2016, a wave originating from a violent cosmic event of the merger of two black holes, which are much more massive objects than planet–star systems and, most importantly, much more energetic. The mass of the new merged black hole was 62 solar masses. Energy equivalent to three solar masses was emitted as gravitational waves. In 2017 astrophysicists also detected a gravitational wave originating from another violent cosmic event—from the merger of two so-called neutron stars that orbited one another.

In summary, when it accounts for gravitational radiation, classical physics does not encounter contradictions with astrophysical observations. There is a long-term (at least millions of years) stability of planets orbiting stars, moons orbiting planets, and, for example, the rings of Saturn (made mostly of water ice) orbiting Saturn, where different rings have different periods of revolution from 5.8 h for the inner edge of the closest ring to 14.3 h for the outer edge of the most distant ring. By the way, there is the following joke by Mark Russel about the rings of Saturn:

> The scientific theory I like best is that the rings of Saturn are composed entirely of lost airline luggage.

However, when it accounts for electromagnetic radiation, classical physics fails to explain the observed relative stability of the everyday world around us and of ourselves. This was the primary (though not the only) motivation for developing a more general theory, called quantum theory or quantum mechanics.

Quantum mechanics employs a different mathematical formalism (a different 'language') compared to classical physics. The quantum language is much more abstract than the classical language, so that the quantum description of phenomena is much more difficult (typically, impossible) to visualize than the classical description. This led to wisecracks like the following:

> To understand something means to derive it from quantum mechanics, which nobody understands.

The mathematical formalism employed by quantum mechanics is called the 'algebra of operators'. Going into details of the algebra of operators is beyond the scope of this book. Let us just state that the primary consequences of the algebra of operators are the so-called *uncertainty relations*. Luckily, the majority of quantum phenomena can be explained qualitatively (and sometimes even quantitatively by the order of magnitude) using just the uncertainty relations.

The primary uncertainty relation involves the coordinate of a particle and the velocity of the particle (later in this chapter, this statement will be formulated more precisely). It could be formulated as follows: the more precisely the coordinate of a particle is known, the less precisely is its velocity known, and the other way around. The commonly accepted notation for the uncertainty of some physical quantity f is Δf, where Δ is the Greek capital letter delta. So, the uncertainty relation between the coordinate of the particle along some x-axis and the projection v_x of the velocity of the particle on the x-axis reads

$$m(\Delta v_x)\Delta x \backsim \hbar, \tag{3.3}$$

where m is the mass of the particle and $\hbar = 1.054 \times 10^{-27}$ g cm^2 s^{-1} is the so-called Planck constant, which is yet another fundamental physical constant. The similar uncertainty relations are between Δv_y and Δy, as well as between Δv_z and Δz.

The everyday objects around us have masses of hundreds, or thousands, or millions of grams. Thus, for everyday objects the product $(\Delta v_x)\Delta x$ is extremely small (because the Planck constant $\hbar$ is extremely small). Therefore, the quantum restriction imposed by equation (3.3) is not observed for everyday objects and they behave according to classical mechanics, which corresponds to the limiting case where the small Planck constant $\hbar$ is neglected ($\hbar$ is approximated by zero). However, for particles such as electrons, whose mass m_e is extremely small ($m_e = 0.911 \times 10^{-27}$ g), from equation (3.3) one obtains $(\Delta v_x)\Delta x \backsim 1$ cm^2 s^{-1}, so that this quantum restriction actually controls the destiny of electrons in atoms.

Incidentally, the uncertainty relations (sometimes called the uncertainty principle) were first introduced by Heisenberg—one of the godfathers of quantum mechanics. Heisenberg and his uncertainty relations are the subject of jokes, such as, the following:

A sign on a railroad station:

These railroads are subject to Heisenberg's uncertainty principle: the position and velocity of a given train cannot be specified at the same time.

There is a physical quantity called *momentum*, p, that typically (though not always!) relates to the velocity as $p = mv$. Therefore, the uncertainty of the momentum relates to the uncertainty of the velocity as $\Delta p = m(\Delta v)$. For example, the relation (3.3) can be represented in the form

$$\Delta p_x \Delta x \backsim \hbar, \tag{3.4}$$

where p_x is the projection of the momentum on the x-axis. Physically, a relation such as (3.4) means that the momentum and the coordinate are not independent of each other, but rather they have a *coupling* (the 'interaction').

A similar uncertainty relation also exists between the energy, E, of a particle and its 'position' t on the time scale

$$\Delta E \; \Delta t \backsim \hbar. \tag{3.5}$$

One of the meanings of the relation (3.5) is that if a particle has a finite lifetime Δt, then it cannot have a definite energy—the uncertainty in its energy is $\Delta E \backsim \hbar/\Delta t$.

Here is another joke concerning Heisenberg, his uncertainty relations (3.5) between the energy and the time, and his uncertainty relation (3.4) between momentum and position:

> Historians have concluded that Heisenberg must have been contemplating his love life when he discovered the uncertainty principle:
> when he had time, he did not have the energy,
> and
> when the moment was right, he could not figure out the position.

One of the most mind-boggling results of quantum mechanics (confirmed by various experiments!) is so-called *tunneling*. In layman's terms, it is as if an object could travel through a wall without making a hole in the wall or damaging the wall in any other way. This turns out to be possible if the object has a relatively small mass and the wall is relatively thin.

As an illustration, let us consider a quantum system, where the effective potential energy $U_{\text{eff}}(r)$ has the shape shown in figure 3.1, where both U_{eff} and r are in arbitrary units. The horizontal dashed line represents one of the possible values of the total energy $E = K + U_{\text{eff}}$. Once again, the fact that the chosen value of the total energy is negative ($E = -3$) should not cause any confusion. This is because the

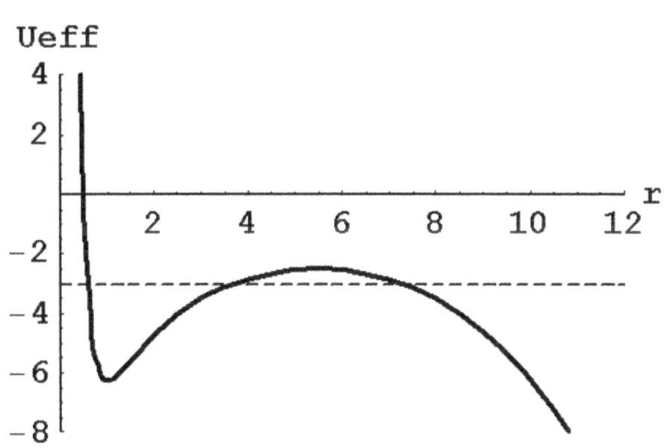

Figure 3.1. Plot of the effective potential energy (solid curve) for some quantum system (both U_{eff} and r are in arbitrary units). The horizontal dashed line represents one of the possible values of the total energy $E = K + U_{\text{eff}}$.

effective potential energy U_{eff} in $E = K + U_{\mathrm{eff}}$ is counted from the zero level chosen to be when a particle would be at the distance from the attractive center $r = 0.5$. Then as the distance r increases from $r = 0.5$, the potential energy decreases and therefore becomes negative, so that the total energy E also could be negative.

A classical particle of energy $E = -3$ could be trapped between the classical turning points $r = 0.58$ and $r = 3.7$. For a particle of energy $E = -3$, the region between $r = 0.58$ and $r = 3.7$ is called the 'classically allowed region'. The particle would not be able to travel to larger distances from the attractive center than $r = 3.7$: it would encounter a barrier (called the 'potential barrier'). Also the classical particle of energy $E = -3$ could travel as an 'almost free' particle in the region $r > 7.3$, but it cannot penetrate the potential barrier and reach distances $r < 7.3$. For the particle of energy $E = -3$, the region of $r > 7.3$ is the second classically allowed region.

The reason why the classical particle cannot be inside the potential barrier is the following. Since the total energy is $E = K + U_{\mathrm{eff}}$, then the kinetic energy $K = E - U_{\mathrm{eff}}$. Inside the barrier the effective potential energy U_{eff} is greater than the total energy E, which corresponds to a negative kinetic energy $K = E - U_{\mathrm{eff}}$. However, from the definition of kinetic energy $K = mv^2/2$, it follows that the kinetic energy cannot be negative—because it is the product of two factors $m/2$ and v^2, neither of which can be negative.

In contrast, a quantum particle, being initially trapped in the region of r between 0.58 and 3.7, can—with some relatively *small probability*—travel under the barrier, reach the other classically allowed region, and move to infinity, thus becoming a free particle. Similarly, a quantum particle located initially in the second classically allowed region of $r > 7.3$, can—with some relatively *small probability*—travel under the barrier and reach the first classically allowed region between $r = 0.58$ and $r = 3.7$ (and then go back).

The physics behind this amazing quantum phenomenon can be understood by using the uncertainty relations. A noticeable probability of tunneling corresponds to the situation where the total energy is just slightly below the top of the potential barrier—as in figure 3.1. This means that a quantum particle spends just a relatively small interval of time Δt under the potential barrier. Since Δt is relatively small, then from the uncertainty relation (3.5) one obtains $\Delta E \backsim \hbar/\Delta t$ to be relatively large. This means that the quantum particle does not have just one definitive value of energy; rather, its energy is somewhere within the band of width $\backsim \hbar/\Delta t$ centered at the most probable value of the energy—such as the band shown in figure 3.2.

Since the width of the energy band is relatively large, then some possible values of the energy within the band are above the top of the potential barrier. This explains why, despite the most probable value of the energy of the quantum particle (shown by the dashed line in figure 3.2) is below the top of the barrier, the particle can be found with some probability on the other side of the barrier.

Tunneling can also be similarly understood by using the other uncertainty relation (3.3). If the most probable value of the total energy is just slightly below the top of the potential barrier—as in in figure 3.2—then the uncertainty of the coordinate of the particle is equal by the order of magnitude to the length Δr of the segment of the dashed line inside the barrier. Since in this situation Δr is relatively

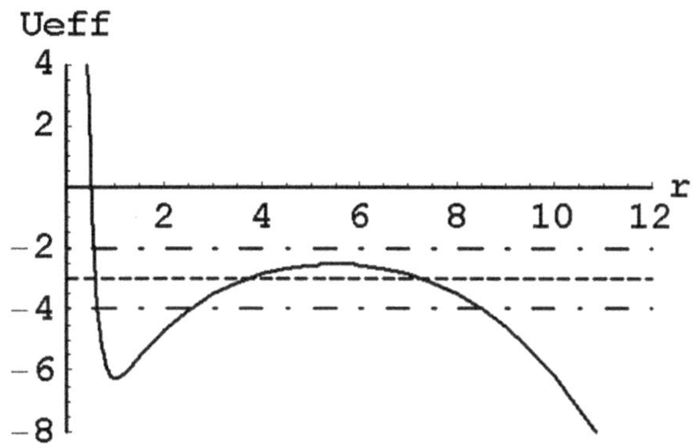

Figure 3.2. Plot of the effective potential energy (solid curve) for some quantum system (both U_{eff} and r are in arbitrary units). Due to the uncertainty relations, the total energy $E = K + U_{\text{eff}}$ is represented by a band; the center of the band is shown by the dashed line, the edges of the band are shown by the dash-dotted lines.

small, then, according to an uncertainty relation similar to equation (3.3), the uncertainty of the particle velocity $\Delta v \sim \hbar/(m\,\Delta r)$ is relatively large. This translates into a relatively large uncertainty of the kinetic energy $K = mv^2/2$; it can be shown that

$$\Delta K \backsim \hbar v/\Delta r. \tag{3.6}$$

The relatively large uncertainty of the kinetic energy translates into a relatively large uncertainty of the total energy. So, this reasoning also leads to the energy band shown in figure 3.2 and to the fact that with some probability the quantum particle can have an energy above the top of the barrier. Thus again, despite the most probable value of the energy of the quantum particle being below the top of the barrier, the particle can be found with some probability on the other side of the barrier.

It should be emphasized that according to equation (3.6), ΔK is proportional to the velocity v of the object under consideration. For example, electrons in atoms and molecules move with relatively large velocities ($\backsim$3 000 000 m s^{-1}), so ΔK is relatively large. However, the objects around us that we encounter in everyday life (macroscopic objects, in contrast to microscopic objects such as atomic and molecular systems) move much more slowly; practically all of them move much more slowly than the speed of sound in air, which is $\backsim$300 m s^{-1}. Therefore for macroscopic objects the uncertainty in the kinetic energy (and therefore the uncertainty in the total energy) is relatively small, so the 'tunneling' of macroscopic objects has never been observed.

So, can a person go through a wall without damaging himself/herself and the wall? The answer is yes, it is *possible*, but extremely *improbable*. Quantum mechanics provides the description of various phenomena in terms of the *probabilities* of different outcomes. The probability of a person going through a wall without damaging himself/herself and the wall is the inverse of an astronomically large number. This does not prevent, however, the existence of jokes like the following:

Question: What is the difference between a quantum mechanic and an auto mechanic?
Answer: A quantum mechanic can get his car into the garage without opening the garage door.

Speaking of probabilities, this concept also has its share of jokes. One example is as follows:

A probability is a desperate attempt by chaos to become stable.

Another example, presented below, seems to be less 'philosophical' than the previous one:

While preparing a patient for surgery, a nurse says: 'This surgery is very risky. Nine out of ten patients die during surgery. But you are lucky: all nine patients before you died during surgery.'

So, how do the uncertainty relations explain the relative stability of atoms, molecules, the world around us, and ourselves? Let us illustrate this by the example of hydrogen atoms that consist of one electron around the proton (the nucleus). The total energy of a hydrogen atom $E = K + U$ is

$$E = m_e v^2/2 - e^2/r, \tag{3.7}$$

where m_e and e are the electron mass and charge, respectively; r is the distance between the electron and the proton. If the momentum p of the electron is related to the electron velocity v by the simple equation $p = m_e v$ (the reason for using the word 'if' in this statement will be explained below), then the total energy can be represented in the form

$$E = p^2/(2m_e) - e^2/r. \tag{3.8}$$

In classical mechanics the coordinate r and the momentum p are *independent* of each other (that is, *not coupled*), so that the momentum can take any value regardless of the coordinate. In particular, classical mechanics (classical equations of the motion) allow the solution where $r = 0$ and $p = 0$, so that the electron is at rest at the location of the proton, which means that the atom has collapsed and is 'dead'.

In contrast, in quantum mechanics the coordinate r and the momentum p are *not independent* of each other, but rather they are *coupled* by an uncertainty relation similar to equation (3.4):

$$\Delta p\, \Delta r \backsim \hbar. \tag{3.9}$$

Therefore, if p is *exactly* equal to zero, meaning that the uncertainty of the momentum $\Delta p = 0$, then the coordinate r cannot be exactly equal to zero, because in this case equation (3.9) yields the infinite uncertainty of the coordinate, $\Delta r = \infty$. The infinite uncertainty of the coordinate means that the electron can be found at any distance from the nucleus, rather than being exactly at $r = 0$. This is how quantum mechanics prohibits the collapse of the atom.

Moreover, using the uncertainty relation (3.9) we can estimate the lowest possible energy of hydrogen atoms, as follows. In the vicinity of the classical solution where $r = 0$ and $p = 0$, we can estimate the total energy in equation (3.8) by substituting p by Δp and r by Δr:

$$E = \Delta p^2/(2m_e) - e^2/\Delta r. \tag{3.10}$$

Let us use equation (3.9) and substitute Δp in equation (3.10) by $\hbar/\Delta r$:

$$E(\Delta r) \backsim [\hbar^2/(2m_e\Delta r^2) - e^2/\Delta r]. \tag{3.11}$$

The plot of the dependence of E on Δr has a shape similar to the shape of $U_{\rm eff}$ in figure 2.2—it has a minimum at some value of Δr. It can be shown that the energy E reaches its minimum at

$$\Delta r = \hbar^2/(m_e e^2). \tag{3.12}$$

After substituting Δr from equation (3.12) in equation (3.11), we obtain the following estimate for the minimum energy:

$$E_{\rm min} \backsim [-m_e e^4/(2\hbar^2)]. \tag{3.13}$$

Actually, the approximate value of $E_{\rm min}$ from equation (3.11) coincides with the exact quantum solution for hydrogen atoms:

$$E_{\rm min} = -m_e e^4/(2\hbar^2). \tag{3.14}$$

This is the *lowest energy state* (called the *ground state*) of hydrogen atoms. Since there are no energy states lower than $E_{\rm min}$, then in this lowest energy state, the electron cannot lose energy (by radiation); if it were to lose some energy, the electron would be in the state of energy $E < E_{\rm min}$, but such states do not exist. This is how quantum mechanics explains the relative stability of atoms, molecules, the world around us, and ourselves. But is this explanation complete? Let us discuss this below.

According to the uncertainty relation (3.9), the electron does not have a *definite* orbit around the nucleus, but rather it can be found at different distances from the nucleus with various probabilities. Therefore, instead of pointing to one specific location of the electron, quantum mechanics describes the situation as an 'electron cloud' around the nucleus—even when speaking of just one electron; in quantum mechanics an electron is not a point-like particle, but rather it is 'spread' over some interval of space.

Similarly to the cloud of electron coordinates, there is a cloud of electron velocities, meaning that there is a probability distribution of velocities of the electron. There is also a cloud of electron accelerations, meaning that there is a probability distribution of accelerations of the electron. Since there is a probability of acceleration, then there is a probability of electromagnetic radiation. (Reminder: a charge moving with acceleration emits electromagnetic radiation.) In the opinion of the author of this book, the only way to reconcile this fact with the stability of the ground state is the following.

For any part of the electron cloud that emits a particular electromagnetic wave there is another part of the electron cloud that emits another electromagnetic wave of the same amplitude, but this other electromagnetic wave has the opposite phase compared to the first one, so that these two waves cancel each other out. This is called 'destructive interference', as illustrated in figure 3.3.

Incidentally, the exact value of the *average* size of the electron cloud in the lowest state of hydrogen atoms actually coincides with the right-hand side of equation (3.12) and is called the Bohr radius a_B:

$$a_B = \hbar^2/(m_e e^2) = 0.529 \times 10^{-8}\ \text{cm}. \tag{3.15}$$

So, the Bohr radius a_B is of the order of 1/10 000 000 of a millimeter, and this is the characteristic size of the ground state of atoms with few electrons.

In atomic physics the energy is typically measured in electron-volts (denoted eV). One eV is the energy acquired by an electric charge while passing the potential difference of one volt in an electric field. In these units, the lowest energy of hydrogen atoms, given by equation (3.14) is −13.6 eV. So, for an electron in the lowest energy state of a hydrogen atom to escape to 'freedom' (to relatively large distances) from the nucleus, it should be given an amount of energy of at least 13.6 eV. When this happens, the atom is said to be 'ionized'. The minimum amount of energy needed to ionize an atom from its ground state is called the 'ionization potential' of the atom. Thus, for hydrogen atoms the ionization potential is 13.6 eV.

Let us bring up a provocative question: was it absolutely necessary to develop quantum mechanics to explain the relative stability of atoms and molecules? Actually, the answer is no! It is possible to modify classical mechanics (to generalize it) in such a way that it explains the relative stability of atoms and molecules, as follows.

In year 2002, Oks and Uzer published a paper in *Journal of Physics B: Atomic, Molecular and Optical Physics* [1] where they provided such a generalization of classical mechanics. The focus of that paper was on the so-called *conserved quantities*.

In general, conserved quantities are those that remain unchanged during a physical process. In particular, in classical mechanics, for the electron orbiting a

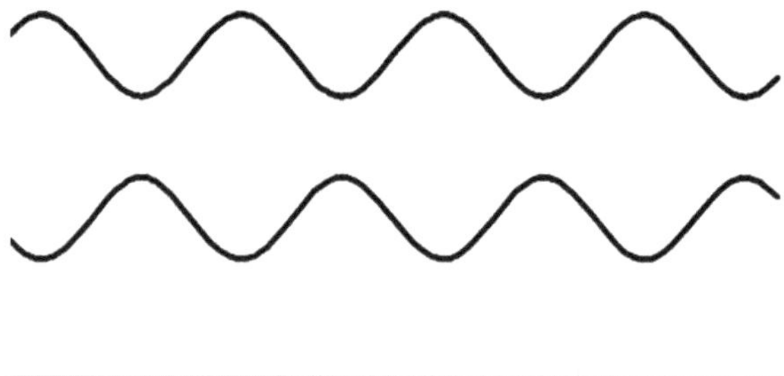

Figure 3.3. Illustration of the destructive interference: two waves of equal amplitude and opposite phase cancel each other out—their sum is zero (straight line, no wave).

nucleus or for a planet orbiting a star, in addition to the energy, there are other conserved quantities as presented below.

One such quantity is the angular momentum **M** (which, if measured in units of the Planck constant $\hbar$ is commonly denoted as **L**). The angular momentum **M** (or **L**) is a *vector*, which means that it is characterized not only by its absolute value, but also by its direction in space. It is commonly accepted to denote vectors in bold type. The coordinate **r** of a particle in space is also a vector (called the radius-vector). The momentum **p** of the particle is also a vector.

The absolute value of the angular momentum is defined through the radius-vector **r** and the momentum **p** of a particle as $\mathbf{M} = \mathbf{rp} \sin \theta$, where θ is the angle between vectors **r** and **p**. The direction of the angular momentum is perpendicular to both vectors **r** and vector **p**. So, while both **r** and **p**, as well as the angle θ, vary as the electron orbits the nucleus or the planet orbits the star, the combination $rp \sin \theta$ remains constant and so does the direction of the angular momentum vector **M** (or **L**). In other words, the angular momentum vector is conserved.

Actually, the angular momentum vector is conserved even in a more general situation when a particle moves in any attractive potential such that the force on the particle depends only on the absolute value r of its radius-vector **r**, but does not depend on the direction of the radius-vector **r**. This class of potentials is called spherically symmetric (or central) potentials.

It is very important that in classical mechanics all central potentials possess yet another conserved vector quantity in addition to the conservation of the angular momentum. (For non-central potentials in classical mechanics, the angular momentum is not conserved, but another conserved quantity or quantities still exist.) For the particular case of an electron orbiting a nucleus or a planet orbiting a star (that is, for the situations where the force on the particle is proportional to $1/r^2$), this additional conserved vector quantity is usually called the Runge–Lenz vector—although it should have been called the Hermann–Bernoulli–Laplace–Runge–Lenz vector, the names of the five scientists who discovered (or re-discovered) this conserved quantity throughout the history of physics.

Figure 3.4 shows the classical elliptical orbit of an electron around a nucleus or of a planet around a star. There are two conserved vector quantities: the angular momentum vector **L** perpendicular to the plane of the orbit and the Runge–Lenz vector **A** along the major axis of the ellipse. Vector **B** is just a vector perpendicular to vectors **L** and **A**: it is not an independent conserved quantity.

The key point of the paper by Oks and Uzer [1] was that in classical mechanics, modified in such a way that all conserved quantities are treated on more or less equal footing, the radius-vector **r** and the momentum **p** of the electron orbiting a nucleus become *coupled*. Therefore, in this generalized classical mechanics, just as in quantum mechanics, the electron cannot fall to the nucleus; there is a state of the lowest energy $E_{\min}$—the ground state. The value of $E_{\min}$ in generalized classical mechanics turns out to exactly coincide with the corresponding quantum value.

But in contrast to quantum mechanics, generalized classical mechanics helps in visualizing what happens when an electron orbiting the nucleus loses energy by

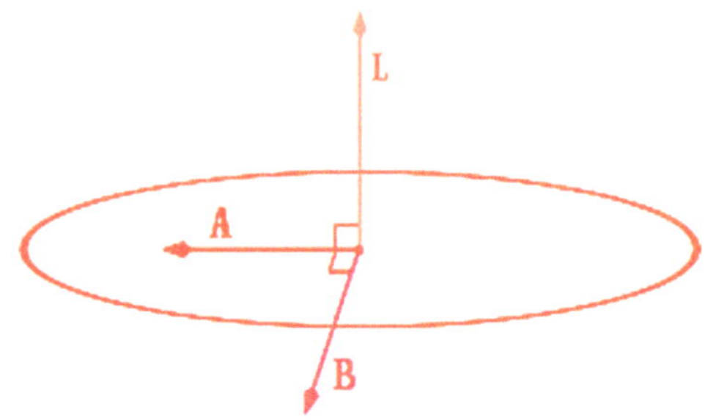

Figure 3.4. Classical elliptical orbit of an electron around a nucleus or of a planet around a star. There are two conserved vector quantities: the angular momentum vector **L** perpendicular to the plane of the orbit and the Runge–Lenz vector **A** along the major axis of the ellipse. Vector **B** is just a vector perpendicular to vectors **L** and **A**—it is not an independent conserved quantity.

emitting electromagnetic radiation. Namely, as the electron gradually loses energy and its energy approaches the value of E_{min}, *time for the electron slows down*! It is a *non-Einsteinian time dilation*. Reminder: in Einstein's special relativity theory, time slows down if objects move with velocities comparable to the speed of light. That is not the case here; the electron orbiting the nucleus moves with a velocity about two orders of magnitude smaller (that is, ~100 times smaller) than the speed of light, so that Einsteinian time dilation is a negligibly small effect.

So, the interpretation of the rigorous results of generalized classical mechanics is that as the energy of the orbiting electron gradually approaches the value of E_{min}, time slows down more and more. As the electron energy reaches the value of E_{min}, for the electron *time stops*! In this ground state the electron has zero velocity and zero acceleration. Since the acceleration is zero, the electron does not emit electromagnetic radiation and therefore does not lose energy any further—the classical atom becomes stable.

Interestingly enough, in such a *radiationless state*, the momentum of the electron is not zero despite its velocity being zero. This is because in generalized classical mechanics, the momentum **p** and the velocity **v** of the electron are no longer connected by the simple relation $\mathbf{p} = m_e\mathbf{v}$: rather, they are connected in a more complicated way. (Incidentally, a more complicated relation between mechanics, the momentum **p** and the velocity **v** is not a feature unique to generalized classical mechanics only; there is an analogous situation in traditional classical mechanics when a charged particle is located in a magnetic field—the particle could have zero velocity, but a non-zero momentum.) The non-zero momentum in this classical ground state indicates that this is not a static situation (even though the electron velocity is zero); there is an *intrinsic dynamics* in this state. Advanced readers can find more details in appendix A.

For microscopic systems, the attractive potential where the force is proportional to $1/r^2$ is called *the Coulomb potential*. Here is a joke that uses a proportionality:

> The chance of a slice of bread falling buttered side down is directly proportional to the cost of the carpet.

Generalized classical mechanics can yield for *non-Coulomb* potentials results similar to those obtained for the Coulomb potential. This is because for the non-Coulomb potentials conserved quantities still exist in traditional classical mechanics (as was mentioned above), so that a description using generalized classical mechanics can still be developed, thus leading to basically the same results as in quantum mechanics, but much easier to visualize. In 2010, Camarena and Oks published a paper in the *International Review of Atomic and Molecular Physics* [2] considering an attractive potential slightly different from the Coulomb potential and treating it in the frame of generalized classical mechanics.

It should be emphasized that for the overwhelming majority of microscopic systems the additional conserved quantities exist only classically and not quantally (with a few exceptions, such as the hydrogen atom). Thus, classical mechanics has an intrinsic, built-in advantage over quantum mechanics in this regard.

Speaking of potentials slightly different from the Coulomb potential, let us discuss the relativistic effects for the electron orbiting the nucleus. The allowance for the relativistic effects makes the potential slightly different from the Coulomb potential.

Figure 3.5 shows the slowly rotating classical elliptical orbit of the electron around a nucleus resulting from the allowance for relativistic corrections (or for another slight deviation from the Coulomb potential). It shows that the elliptical

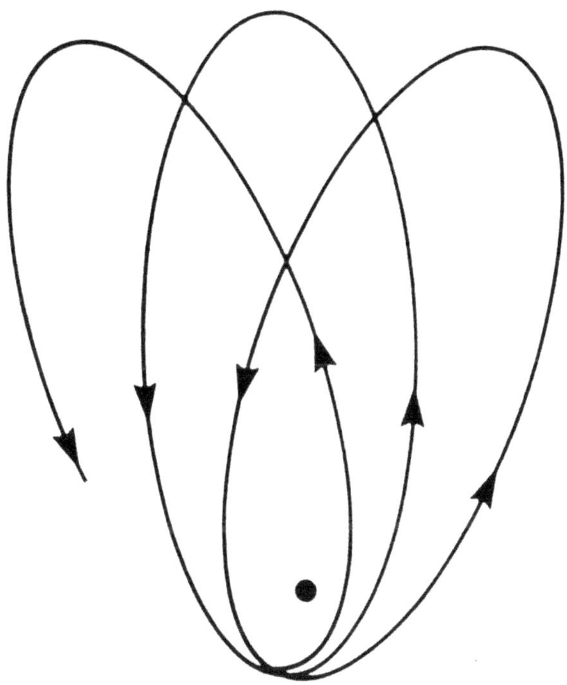

Figure 3.5. The slowly rotating (precessing) classical elliptical orbit of the electron around a nucleus resulting from the allowance for relativistic corrections (or for another slight deviation from the inverse square ($1/r^2$) force law, that is from the Coulomb potential).

orbit slowly rotates in the plane of the orbit; such a slow rotation of the orbit is called *precession*.

The precession of the orbit is slow (compared to the motion of the electron along the ellipse) as long as the relativistic corrections are relatively small. The smallness of the relativistic corrections is controlled by the factor $[Ze^2/(\hbar c)]^2$, where Z is the nuclear charge (in units of the absolute value e of the electron charge). The dimensionless combination $e^2/(\hbar c)$ is one of the most fundamental dimensionless constants in physics. It is called the *fine structure constant* and numerically it is very close to 1/137. So, for relatively small nuclear charges ($Z \backsim 10$ or smaller), the factor $[Ze^2/(\hbar c)]^2$ is $\backsim$ 1/100 or much smaller.

Incidentally, the relative closeness of the inverse fine structure constant $\hbar c/e^2$ to the integer number 137 motivated many attempts by different physicists to 'deduce' the 'magic' number 137 from various considerations. More details of this will be discussed in chapter 5.

References

[1] Oks E and Uzer T 2002 *J. Phys. B: At. Mol. Opt. Phys.* **35** 165

[2] Camarena J and Oks E 2010 *Int. Rev. At. Mol. Phys.* **1** 143

IOP Publishing

Unexpected Similarities of the Universe with Atomic and Molecular Systems: What a Beautiful World

Eugene Oks

Chapter 4

Crossing of the micro- and macro-worlds: similarities between planet–star or satellite–planet systems and giant (Rydberg) atoms with a diameter twice the size of a human hair

When new experimental facts motivate physicists to develop a more general theory, the previous theory usually does not get put into the trash. Rather, a more general theory embraces the previous theory as some limiting case. For example, Einstein's special relativity theory embraces Newtonian mechanics as a limiting case where an object moves with velocities much smaller than the speed of light. Similarly, quantum physics embraces classical physics as some limiting case illustrated below by the example of hydrogen atoms.

The energy of the lowest (ground) state of hydrogen atoms is given by equation (3.14): $E_{\min} = -m_e e^4/(2\hbar^2)$. According to quantum mechanics, this is just one of many so-called 'stationary states' of the energies

$$E_n = -m_e e^4/(2n^2\hbar^2), \tag{4.1}$$

where $n = 1, 2, 3, \ldots$ is any positive integer and it is called the *principal quantum number*. The rigorous concept of stationary states originates from the mathematical formalism of quantum mechanics (from the algebra of operators). The physical consequence is that the electron in hydrogen atoms can have only energies given by equation (4.1), but no energies between those selected values. This statement can be mathematically formulated as follows: the electron in hydrogen atoms can have only a *discrete set* of energies, rather than a *continuous set* of energies. (The same is true for any atom or molecule, but the equation for the discrete set of energies would differ from equation (4.1).)

doi:10.1088/2053-2563/aafed9ch4

The state of $n = 1$ is the ground state. The state of $n = 2$ is called the first excited state; its energy is higher than that of the ground state. The state of $n = 3$ is called the second excited state; its energy is higher than that of the first excited state, and so on.

This 'ladder of energies' is illustrated in figure 4.1. It shows graphically the positions of the first six energy levels of the hydrogen atom $n = 1, 2, 3, 4, 5, 6$. The top bar represents the energy $E = 0$ (corresponding to infinitely large n)—above this level the electron can go away from the nucleus to arbitrarily large distances; this situation is called ionization of the atom.

According to quantum mechanics, when the electron is in one of the excited states, it can 'spontaneously leap' down to one of the lower energy states, while the energy difference between its initial and final states will be emitted as electromagnetic radiation. These transitions are called 'quantum leaps'.

Concerning the reason why this occurs 'spontaneously', the quantum explanation involves a highly counterintuitive concept, as follows. In the quantum description, the electromagnetic field is equivalent to a quantum system characterized by the potential energy of the parabolic shape shown in figure 4.2. In this kind of potential, the quantum system has the ladder of energy levels, shown in figure 4.2 by horizontal lines for the first four states. The most important is that the ground state ($n = 0$) has an energy still above the bottom of this potential well.

This is again a consequence of the uncertainty relation given by equation (3.4): $\Delta p_x \Delta x \sim \hbar$. Indeed, if the system were to rest at the bottom of the potential well, this

Figure 4.1. Positions of the first six energy levels of the hydrogen atom $n = 1, 2, 3, 4, 5, 6$. The top bar represents the energy $E = 0$ (corresponding to infinitely large n).

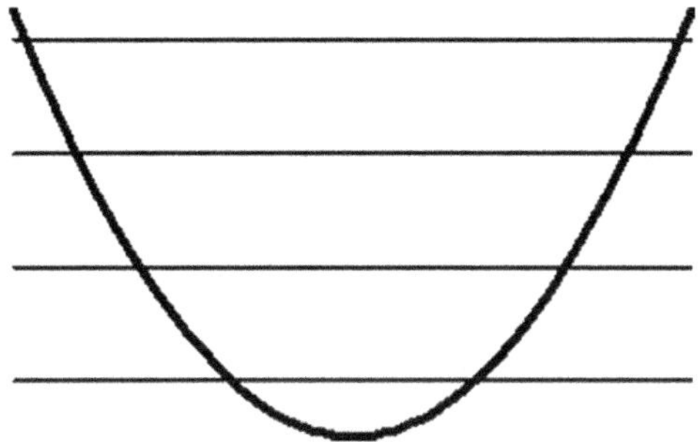

Figure 4.2. Energy levels (horizontal lines) of a quantum system characterized by the potential energy of the parabolic shape (thick curve), the quantum system representing the electromagnetic field in a vacuum.

would mean $\Delta x = 0$. Then $\Delta p_x \sim \hbar/\Delta x$ would be infinite, so that the system could not be at rest at the bottom of the well and thus having p_x exactly equal to zero.

Physically this means that even in the state of the lowest energy, there still exists some electromagnetic field because it has a non-zero energy. This energy is called the 'zero-point energy' (confusingly enough). What we would casually call an empty space (that is, the space where there is no matter), is called a 'vacuum' in physics. So, a vacuum is not actually empty; it is filled with 'zero-point' oscillations (that is, variations in time) of electric and magnetic fields.

From this point of view, even in a model where an atom is isolated from the rest of the world, it cannot be isolated from the zero-point electromagnetic field of the vacuum. So, when the electron in the atom is in one of the excited states, due to the interaction with the vacuum (with the zero-point electromagnetic field) it leaps to one of the lower energy levels and emits the energy difference as electromagnetic radiation. While this is called 'spontaneous emission', it is not spontaneous from the above point of view, rather it is triggered by the interaction with the vacuum.

Incidentally, generalized classical mechanics (discussed in chapter 3 and presented in more detail in appendix A) also yields a set of radiationless energy states given by exactly the same equation as (4.1): $E_n = -m_e e^4/(2n^2\hbar^2)$. In any such state the electron is in equilibrium. However, except for the state of $n = 1$, in all other states the equilibrium is unstable; under any fluctuation the electron starts going *continuously* to lower energies while emitting electromagnetic radiation—until it reaches a lower radiationless state. Then under any fluctuation the process repeats, until the electron reaches the lowest radiationless state.

Now let us get back to quantum mechanics and discuss the average size of the exited states of hydrogen atoms. In the state of the principal quantum number n (having the energy $E_n = -m_e e^4/(2n^2\hbar^2)$),), the average radius r_{aver} of the electron cloud is

$$r_{aver} = n^2 a_{B,} \tag{4.2}$$

where $a_B = \hbar^2/(m_e\, e^2) = 0.529 \times 10^{-8}$ cm is the Bohr radius (as given previously by equation (3.15). Therefore, as n increases, so does the size of the electron cloud. Actually, for sufficiently large n (which is commonly denoted as $n \gg 1$), equation (4.2) is valid not only for hydrogen atoms, but for *any* atoms. Atoms in the states of $n \gg 1$ are called Rydberg atoms.

In the limit of $n \gg 1$, the quantum description practically coincides with the classical description. Therefore, in the states of the principal quantum number $n \gg 1$, the description in terms of the classical elliptical orbit of the electron around the nucleus becomes appropriate.

In the Universe there are astrophysical objects (called H II regions) where hydrogen atoms exist in the states of the principal quantum number $n \sim 1000$ (!). This means that in such states, the diameter of the electron orbit is $\sim 2r_{\mathrm{aver}} \sim 0.01$ cm = 0.1 mm. In other words, these giant atoms have a diameter twice the size of a human hair, the diameter of which is $\sim$ 0.05 mm. This is indeed where the micro-world and our everyday world cross!

We know about the existence of these giant atoms because they emit electromagnetic radiation in the radio range of frequencies (that is, in the range of frequencies close to the frequencies of radio-stations). Actually, like any atom, they emit electromagnetic radiation at a very specific (discrete) set of frequencies observed as a set of so-called *spectral lines*.

A spectral line is a bright line (a discrete line) observed at the background of an otherwise uniform continuous spectrum. Spectral lines are like the 'fingerprints' of atoms and molecules, thus they are used to identify the atomic and molecular components of stars and planets.

The spectral lines emitted by the giant atoms in the H II regions of the Universe are called radio lines, because they are emitted in the radiofrequency range. The radio lines are observed by radio telescopes.

Optical telescopes collect the electromagnetic radiation of visible light, amplify it, and make it possible for the data to be analyzed by instruments called spectrometers. Similarly, radio telescopes collect weak electromagnetic radiation from distant astrophysical objects, amplify it, and make it possible to analyze the data.

Because radio wavelengths are much longer than those of visible light, radio telescopes are very large structures to achieve a resolution comparable to that of optical telescopes. Figure 4.3 shows, as an example, a radio telescope in Ukraine.

Figure 4.3. A radio telescope in Hrakovo, Ukraine.

Speaking of arrays of antennas, of which radio telescopes consist, the following joke comes to mind:

> Two antennas decided to get married. The wedding was terrible, but the reception was great.

Radio telescopes are not used exclusively for registering radio lines from H II regions. They are also used for studying electromagnetic radiation in the radio-frequency range from stars, galaxies, black holes, and other astrophysical objects.

Concluding this chapter, it is appropriate to quote the famous physicist Richard Feynman [1]:

> Astronomy is older than physics. In fact, it got physics started by showing the beautiful simplicity of the motion of the stars and planets, the understanding of which was the beginning of physics. But the most remarkable discovery in all of astronomy is that the stars are made of atoms of the same kind as those on the Earth.

Reference

[1] Feynman R 2015 *The Feynman Lectures on Physics* vol 1 (New York: Basic Books) ch 3

IOP Publishing

Unexpected Similarities of the Universe with Atomic and Molecular Systems: What a Beautiful World

Eugene Oks

Chapter 5

The magic of numbers: a fine-tuned Universe?

In chapter 3 it was mentioned that the relative closeness of the inverse fine structure constant $\hbar c/e^2$ (which is one of the few fundamental dimensionless constants in nature) to the integer number 137, motivated many attempts by different physicists to 'deduce' the 'magic' number 137 from various considerations. Before going into detail on their attempts, it should be emphasized that physically, the probability that an electron will absorb a photon (which is the carrier of the electromagnetic interaction) is the inverse of this number, that is, 1/137. Obviously, this number can be considered as an 'overlap' of quantum mechanics (represented by the Planck constant $\hbar$), Einstein's relativity (represented by the speed of light c), and electromagnetism (represented by the electron charge e).

This number 137 also establishes the limit for the nuclear charge Z (in units of e) of various atoms, namely $Z < 137$. As Z approaches 137, electron velocities in atoms approach the speed of light; so, according to Einstein's relativity, electrons become 'much heavier' to such an extent that the nucleus cannot confine them. Thus this number controls a possible variety of atoms in the Universe.

The first attempt to 'deduce' this was made by astrophysicist Arthur Eddington in 1929 [1]. His starting point was that in the relativistic quantum mechanics developed by Dirac, one electron is described by four functions (called ψ-functions), so a system of two electrons is described by all possible combinations of the ψ_1-functions (characterizing the first electron) with the ψ_2-functions (characterizing the second electron), altogether $4 \times 4 = 16$ combinations. Then Eddington considered the process of a 'rotation' that would interchange these two electrons. He argued that such a process would be described by a matrix of size 16×16, such as the one shown below.

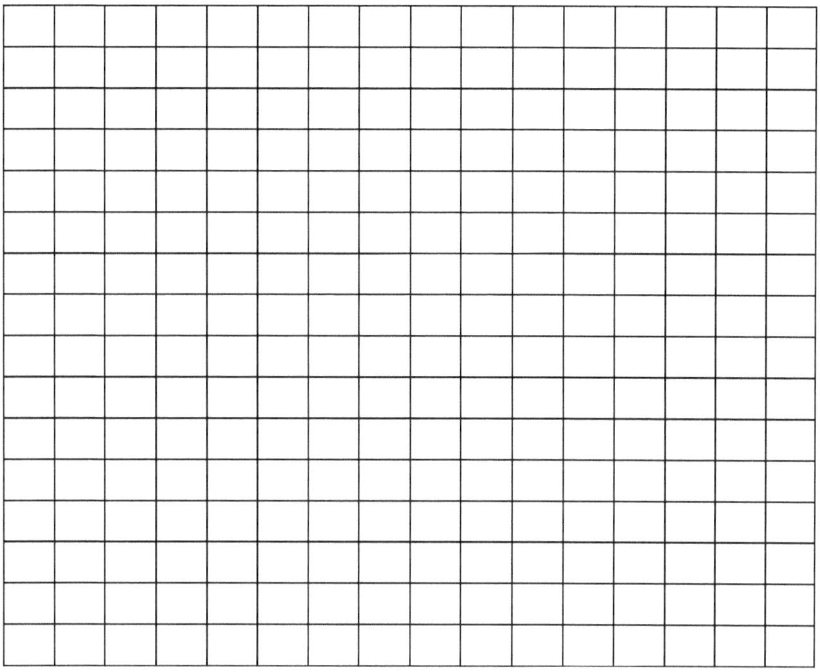

The elements of the matrix can be denoted as a_{ik}, where $i = 1, 2, 3, \ldots, 16$ and $k = 1, 2, 3, \ldots, 16$. In any pair of elements obtained by the interchange of the two suffixes, such as, a_{12} and a_{21}, or a_{35} and a_{53}, the two elements within the pair are not independent of each other. For example, knowing the value of the element a_{12}, it is easy to obtain the value of the element a_{21}.

So, Eddington suggested counting the number of independent elements in this 16×16 matrix. There are 16 elements on the diagonal of the matrix (and they are independent). So, the number of nondiagonal elements is $256 - 16 = 240$. Only one half of the nondiagonal elements is independent, namely $240/2 = 120$. Thus, the total number of independent elements is $120 + 16$ (on the diagonal) $= 136$, which is pretty close to 137.

Eddington's logic might look complicated. There are simpler ways to arrive at the magic number 137.

Arthur Miller in his book *137: Jung, Pauli, and the Pursuit of a Scientific Obsession* [2] wrote that in ancient Hebrew, numbers were written with letters, so that to each letter of the Hebrew alphabet there corresponds a number. The word Kabbalah is written in Hebrew using four letters, whose corresponding numbers are 5, 30, 2, and 100. It is amazing that the sum of these four numbers is equal to 137 (!).

I can offer my own exercises in 'deducing' the number 137. One option is to represent 137 as follows:

$$1 + 2^3 + 2^7 = 137$$

Here the appearance of 1 is justified because it is the starting point in the set of positive integers. The appearance of 2 is justified because it is the basis of computer technology. Computers consist of numerous cells, where each cell is in one of two states only; therefore, computer memory and storage capacity is measured in powers of 2 (and, therefore, both 1 and 2 should be good starting points for the initial—symbolic—communication with an extraterrestrial intelligence, if any.)

As for the numbers 3 and 7, they represent the dimensions of the only two types of worlds that survive in the scenario (discussed in detail in chapter 2) where worlds of all kinds of dimensions were born simultaneously and then start colliding with each other.

So, the above option combines both mathematical and physical considerations. The option below is purely mathematical, as follows.

Readers might be familiar with so-called *prime numbers*. As a reminder: a prime number is defined as an integer that has two and only two dividers—namely, it can be divided only by itself and by 1. It should be noted that according to this definition, 1 is not a prime number because it has only one divider: the set of prime numbers starts as 2, 3, 5, 7, 11, 13, 17, 19, and so on.

Probably less familiar to most readers is a more exotic set of prime numbers called the *Stern prime numbers*, named after mathematician Moritz Stern. The definition of Stern prime numbers is more complicated than for the (usual) prime numbers. An integer is a Stern prime number if it is not the sum of a smaller prime and twice the square of a non-zero integer. (For readers who are not scared of algebra: an integer q is the Stern prime number if there is no smaller prime p and non-zero integer b such that $q = p + 2b^2$.) The known Stern prime numbers are

$$2, 3, 17, 137, 227, 977, 1187, 1493.$$

For example, if we try subtracting from 137 the first few squares doubled—in order—we obtain {135, 129, 119, 105, 87, 65, 39, 9}, none of which are prime. That means that 137 is a Stern prime number.

But what singles out 137 from the above sequence of Stern prime numbers? The author of this book brings to the reader's attention that 137 is the *fourth* Stern prime number and that four is the dimension of our space–time (three spatial dimensions plus one dimension of time).

On the one hand, it might seem that the attempts to 'deduce' 137 should not be taken seriously because more accurate measurements showed that the fundamental dimensionless constant $\hbar c/e^2$ is not exactly equal to 137. Rather, it turned out that $\hbar c/e^2$ is approximately equal to 137.035 999.

There is the following joke about Wolfgang Pauli, one of the godfathers of quantum mechanics (the joke is also partially reproduced by David Gross in [3]). It goes like this:

> Pauli arrives in Heaven and is given, as befits a theoretical physicist, an appointment with God. When granted the customary free wish, he requests that God explain to him why $\hbar c/e^2$ is approximately equal to 137.035 999. God goes to the blackboard and starts writing furiously. Pauli watches with pleasure, but soon starts violently shaking his head in disagreement—indicating that God made an error.

Incidentally, physicist Keith Partain proposed a combination of numbers that is very close to 137.035 999. Namely, he proposed

$$4\pi^3 + \pi^2 + \pi = 137.036\,304.$$

Here π is the ratio of the perimeter of any circle to its diameter; π is approximately equal to 3.141 592 653 589 793 238 462 643 383 279 502 884 (it should be another good starting point for the initial—symbolic—communication with an extraterrestrial intelligence, if any). Partain did not explain why the numbers 2, 3, and 4 enter his equation. One could guess that number 2 enters here because it is the basis of computer technology, as mentioned above; number 3 represents the number of spatial dimensions of our world, while number 4 represents the four-dimensional space–time.

On the other hand, all the attempts to 'deduce' the *integer* 137 were not just exercises of human spirit. Indeed, the possible variety of atoms in the Universe (atoms of different nuclear charges) is *exactly equal* to 137. So, perhaps some of the symbolism behind deducing the integer 137 might make sense.

Incidentally, the author of this book encountered a symbolism in numbers while visiting the Italian city of Ravenna. It is famous for the mosaic masterpieces in its temples built in the 5th and 6th centuries—in particular, in the Mausoleum of Galla Placidia, built by Galla Placidia, the daughter of Emperor Theodosius I.

The walls and the ceiling of this mausoleum are covered in very beautiful mosaics, as partially shown in figure 5.1. The ceiling shows numerous stars.

As the local tour guide brought our group inside the mausoleum under this ceiling, she told us that there are 327 stars on the ceiling. She noted to us that when such temples or mausoleums were built, there was always some religious and/or astronomical symbolism hidden in numbers. Then she said that the number of stars on the ceiling was chosen to be 327 because this number can be divided by 3, with the number 3 symbolizing the Holy Trinity. I could not help asking her the following question: 'There is an infinite number of integers that can be divided by 3. So, why 327?' She responded that historians and archeologists tried to figure this out, but failed.

As the tour continued to other rooms of the Mausoleum, I followed the tour, but simultaneously was thinking about the number 327. I thought: 'After dividing 327 by 3, one gets 109. What kind of religious or astronomical symbolism could be hidden in number 109? Obviously, 109 is a prime number (it can be divided only by itself and by 1), but what could distinguish 109 out of the infinite set of prime numbers?' Then I came up with an idea to calculate in my mind (while still following the tour), what is the place of number 109 in the sequence of prime numbers (that starts from number 2). It turned out that 109 is the 29th prime number. 'Eureka!' (I shouted this in my mind like the ancient Greek scholar Archimedes after he had stepped into a bath and noticed that the water level rose, whereupon he suddenly understood that the volume of water displaced must be equal to the volume of the part of his body he had submerged.) 'Eureka!' 29 is the number of days in the lunar month, that is, from one new moon to the next (called the 'synodic month'). So, indeed, 327 stars on the ceiling combined both religious and astronomical symbolism.

Figure 5.1. Mosaics in the Mausoleum of Galla Placidia in Ravenna, Italy, showing the ceiling with numerous stars.

At the end of the tour I told the tour guide about my 'discovery'. She was stunned! Then she wrote down my explanation word-by-word. Finally, she said that from now on, every tour guide bringing tourists to the Mausoleum of Galla Placidia would tell them my explanation and refer to me. I am not sure that tour guides in Ravenna now actually do this (nor that I need this 'glory'), but this episode, of course, enhanced my impressions of Ravenna.

There are other dimensionless fundamental constants in nature in addition to the number 137. Another is the number of spatial dimensions in our Universe $D = 3$, discussed in detail in chapter 2.

Yet another dimensionless fundamental constant is $\varepsilon = 0.007$ describing the strength of the force that binds atomic nuclei together. For example, as two protons and the neutrons undergo fusion into the nucleus of a helium atom, a 0.007 fraction of the total mass of the two protons and two neutrons is released, basically as heat. This number corresponds to the strength of the nuclear force. Incidentally, this is how our Sun works as a 'power station'; the hydrogen nucleus (which is a proton) and its heavier counterpart deuterium (whose nucleus has one proton and one neutron) supply two protons and two neutrons that convert 0.007 of their mass into energy while fusing into helium.

Astronomer Martin Rees, in his book *Just Six Numbers* [4], noted, in particular, that if this number were only a bit smaller—0.006 instead of 0.007—a proton could not bond to a neutron, and the Universe would consist only of hydrogen. The consequence: no chemistry, no life. And if this number were slightly larger, just 0.008, fusion would be so ready and rapid that no hydrogen would have survived the Big Bang. The consequence: no solar systems, no life.

Another example of a fundamental dimensionless constant discussed by Rees is $Q = 1/100\ 000$. It represents the relative amplitude of ripples in the expanding universe that seed the growth of such structures as planets and galaxies. If this number were smaller, the Universe would be just a cloud of cold gas: no life. If it were larger, big blobs of matter would have consolidated into giant black holes. Such a Universe would be so stormy that stars or planetary systems could not survive.

There is also a dimensional quantity, whose value is crucial for the existence of planetary systems and life. It has to do with the so-called *dark energy* that produces the anti-gravitation effect; it is responsible for the expansion of our Universe. It is called dark energy because so far physicists do not know what it actually is—despite it constituting about 70% of the energy density of our Universe. If there were slightly more dark energy, the Universe would expand so rapidly that the formation of galaxies would become impossible. No galaxies means no structures such as planetary systems around stars, and consequently no life.

All of the above led to the concept of the *fine-tuned Universe*, meaning that if any of several fundamental constants had been only a tiny bit different, there would be no matter, no diversity of elements, and thus no life. Given this situation, some might revert to theology. However, physicists entertained another idea, called the *multiverse*.

The multiverse is the idea that a large (possibly, infinite) array of universes erupted from a large (possibly, infinite) array of 'big bangs'. Each of these universes is controlled by a different set of fundamental constants. Our Universe is just a particular one from this array, such that the fundamental constants controlling it allow the possibility of life. Rees illustrated this as follows [5]: 'The analogy here is of a ready-made clothes shop. If there is a large stock of clothing, you're not surprised to find a suit that fits.'

An interesting comment on the concept of the fine-tuned Universe was made by Tiberiu Tesileanu, a physicist working in neuroscience [6]. He wrote: 'The fact that changing the laws of physics a little bit would not allow for life does not imply that larger changes also would not. As an analogy, put a human in water and he will die; instead, give gills to him and he will also die; but give him gills *and* put him in water and he will be fine.'

A recent paper by Howe, Grohs, and Adams published in 2018 in *Physical Review* D [7] has provided a rigorous physical example in the spirit of the above comment. The authors considered an alternative Universe where the so-called weak force (or weak interaction), which is one of the four known fundamental forces in nature, is less efficient (that is, the weak force is 'weaker') than in our Universe. They demonstrated that such a Universe could still support life if another fundamental constant were to simultaneously decrease. This other fundamental constant is the ratio of so-called baryons (which include protons and neutrons) to photons (the quanta of light).

Thus, in combination with the concept of the multiverse, this means that there could be a group of several (or even many) universes, each controlled by a set of fundamental constants very significantly different from the corresponding sets in other universes from the same group, and yet all of them could support life. However, the life in each of these universes would be very much unlike the life in the other universes from the same group.

References

[1] Eddington A S 1929 *Proc. R. Soc. Lond.* A **122** 358

[2] Miller A I 2009 *137: Jung, Pauli, and the Pursuit of a Scientific Obsession* (New York: Norton)

[3] Gross D J 1989 *Phys. Today* **42** 9

[4] Rees M 2000 *Just Six Numbers* (New York: Basic)

[5] Lemley B 2000 Why there is life *Discover* Novermber 2000 issue

[6] Tesileanu T 2015 Is it true that life would not exist if we changed any of the fundamental constants for a tiny bit? https://www.quora.com/Is-it-true-that-life-would-not-exist-if-we-changed-any-of-the-fundamental-constants-for-a-tiny-bit

[7] Howe A R, Grohs E and Adams F C 2018 *Phys. Rev.* D **98** 063014

Chapter 6

Similarities between the motion of a satellite with constant acceleration and the Stark effect in Rydberg atoms

Our Earth has a natural satellite—the Moon, as everybody knows. Just like a planet orbiting a star moves with an acceleration caused by the gravitational force of the star, the Moon orbits the Earth with an acceleration caused by the gravitational force of the Earth. Artificial satellites that do not have their own propulsion (that is, do not have their own rocket engine) after being put into orbit around the Earth, also orbit the Earth with an acceleration caused by the gravitational force of the Earth.

However, there are situations where in addition to gravitational acceleration, a satellite also has a constant acceleration in some direction. For example, the additional constant acceleration could be due to the relatively small thrust of the satellite's rocket engine.

Interestingly enough, the satellite could be under additional constant acceleration even without a rocket engine; the acceleration would be due to the pressure of solar light. (More details on the latter can be found in section 3.5 of Beletsky's book *Essays on the Motion of Celestial Bodies* [1].)

Incidentally, the motion due to the pressure of light is called *lightsailing*. The source of light could be a star or a laser. Lightsail technology could be practically used for the transportation of goods between planets or between stars. Some lightsails have already been designed and constructed. The first spacecraft that successfully demonstrated solar sail technology in interplanetary space was the Japanese IKAROS (Interplanetary Kite-craft Accelerated by Radiation Of the Sun). It was launched in May 2010 and it passed Venus in December 2010, thus successfully completing the mission.

doi:10.1088/2053-2563/aafed9ch6

Another lightsail project, called Breakthrough Starshot, proposed in the USA is more challenging. The goal is for this lightsail to travel to our nearest star Proxima Centauri and to its Earth-sized planet Proxima Centauri b, which is considered capable of supporting life. This lightsail, whose launch is tentatively scheduled around the year 2036, would travel at a speed of 15%–20% of the speed of light. Sending it would require a multi-kilometer array of beam-steerable lasers. The journey would take about 20 or 30 years. Then it would take another approximately four years for the message from the lightsail to reach Earth.

The latest mind-boggling news about lightsails is the following. In October 2017 astronomers in Hawaii discovered the first ever *intergalactic* object traveling through our solar system. They named it Oumuamua, which means 'messenger from afar arriving first' or 'first distant messenger' in Hawaiian. It has a needle-like, roughly oval form about 400 m by 40 m in size.

Based on the unusual shape and an unusual acceleration of Oumuamua, Shmuel Bialu and Abraham Loeb of the Harvard Smithsonian Center for Astrophysics pointed out that Oumuamua could have been a 'fully operational probe sent intentionally to Earth vicinity by an alien civilization'. Bialu and Loeb introduced this result in a paper published in November 2018 in *Astrophysical Journal Letters* [2]. We really live in very interesting times!

It is yet unknown how long Oumuamua has been traveling among the stars. It has been speculated that Oumuamua has traveled for several tens of millions of years by now. It is estimated that Oumuamua will leave our solar system in 2022 by passing Neptune's orbit and then traveling to interstellar space in the direction of the constellation Pegasus.

Now let us come back to the motion of a satellite under additional constant acceleration. The most amazing thing about the motion of such a satellite is that it has a direct analogy in the micro-world. The analogy is the Stark effect in Rydberg atoms.

First, let us recall that we discussed Rydberg atoms in chapter 4. These are atoms in the states of the principal quantum number $n \gg 1$ (also called highly excited atoms). In the current chapter we consider one-electron Rydberg atoms, that is, highly excited hydrogen atoms or highly excited hydrogen-like ions.

Second, in chapter 4 we also briefly discussed spectral lines, resulting from quantum transitions between the energy levels, and mentioned that spectral lines are like 'fingerprints' of atoms and molecules, such that they are used to identify the atomic and molecular components of stars and planets. Now we will introduce the Stark effect.

In 1913 the German physicist Johannes Stark discovered that a hydrogen spectral line splits into several components under an applied static electric field and that the separation between the components is linearly proportional to the electric field, in the case where the electric field is relatively small. In 1919 he was awarded the Nobel Prize for this discovery. In the same year, 1913, this effect was also discovered independently by the Italian physicist Antonio Lo Surdo (but he was not awarded the Nobel Prize).

Figure 6.1 shows, as an example, the observed Stark splitting of the hydrogen Balmer-alpha line, that is, of the spectral line corresponding to the quantum radiative transition from the hydrogen energy level of $n = 3$ to the hydrogen energy

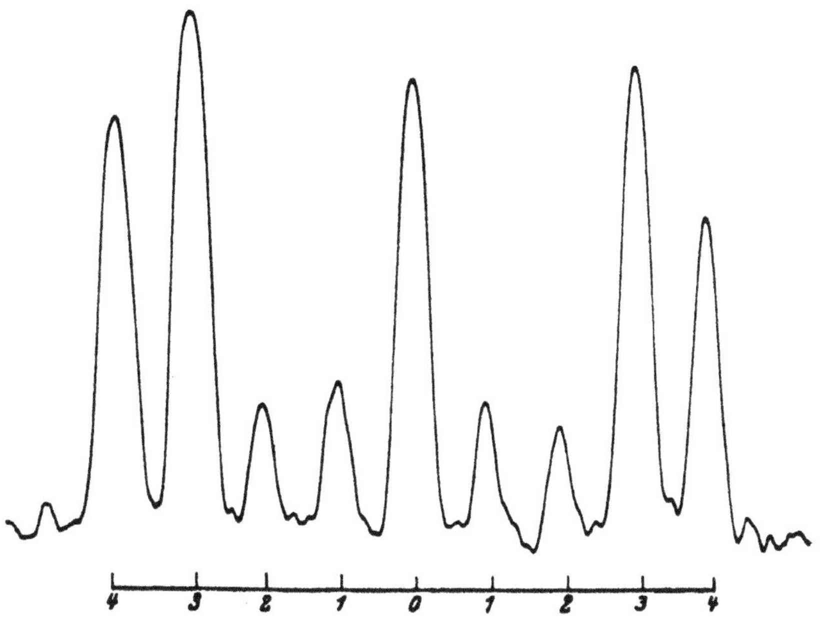

Figure 6.1. Stark splitting of the hydrogen Balmer-alpha line observed by Mark and Wierl [3]. The unperturbed position of this spectral line is indicated by 0.

level of $n = 2$. This is one of the experimental results published by Mark and Wierl in 1929 in a series of articles in *Zeitschrift für Physik* [3]. The hydrogen Balmer-alpha line (usually denoted as H_α) happens to be the most intense spectral line in the visible spectrum of the Sun.

The Stark splitting of hydrogen spectral lines occurs because each energy level (depicted, for example, in figure 4.1) undergoes Stark splitting into an equidistant 'ladder' of sublevels. For a relatively small external electric field F, the separation between the sublevels (which are called Stark sublevels) is linearly proportional to F. The observed Stark splitting of a particular hydrogen spectral line corresponds to a set of quantum transitions between the Stark sublevels of the upper and lower energy levels involved in the radiative transition.

I can offer the following analogy. When you press some key on a piano or on a musical keyboard, it produces a sound of just one tonality. What if, under some change of the conditions in the room, as you press the same key you were to hear a chord of several tonalities? Would you not be surprised?

A more rigorous analogy would be to speak about a tuning fork rather than a piano key. This is because the tuning fork, being struck, produces a sound of just one specific frequency (one specific pitch). In contrast, by pressing a piano key one produces a sound comprised of one specific frequency (called the fundamental frequency) and of so-called overtones, whose frequencies are integer multiples of the fundamental frequency. So, imagine that you strike a tuning fork, but instead of the sound of just one specific pitch you were to hear a chord of several pitches, whose frequencies are relatively close to each other.

This is a better analogy for the Stark effect for the following reason. In the absence of an external electric field, a spectral line has one specific frequency—just like the sound of a tuning fork. Under an applied static electric field, the same spectral line becomes a set of several relatively close frequencies, whose separation from each other is controlled by the external field (the separation being linearly proportional to the external field, when the latter is relatively small).

While experiencing this effect with a tuning fork or with a piano key is just a fantasy, and even if it were not a fantasy would lead only to an amusement, the Stark effect in the course of time led to very real, practically important applications. Most applications are in the area of plasma diagnostics.

Now, what is plasma? Plasma is the fourth fundamental state of matter—in addition to solids, liquids, and gases. Plasma can be thought of as an ionized gas, that is, a gas where atoms (or molecules) have lost electrons and became positive ions surrounding by free electrons. Plasma is the most abundant form of ordinary matter in the Universe (ordinary matter should be distinguished from so-called dark matter, briefly mentioned in chapter 1, whose properties are unknown). Astrophysical examples are (but are not limited to):

- stars and
- intergalactic, interstellar, and interplanetary media, which are likely to be in the low-density plasma state.

Part of the Earth's atmosphere (called the ionosphere) is in the plasma state. Lightning and polar aurorae are also plasmas.

In addition to the above examples of natural plasmas, various types of plasmas are artificially produced in laboratories. Examples are (but are not limited to) magnetically controlled or laser-controlled thermonuclear fusion, plasma-based lasers (in the infrared, visible, ultraviolet, and x-ray ranges), and technological discharges for plasma chemistry (including plasma surface processing for manufacturing microchips and for nano-technologies).

The most practically important plasma research focuses on controlled thermonuclear fusion. Thermonuclear fusion is the process that fuels stars, including the Sun. The goal of research into controlled thermonuclear fusion is to provide a virtually inexhaustible source of energy—the 'fuel' would be ocean water and the process would have a high power output per liter.

A crucial part of plasma research and its practically important applications is *plasma diagnostics*, i.e. a suite of methods for the experimental determination of a variety of plasma parameters and the parameters of various electric and magnetic fields in plasmas. In other areas of physics the corresponding suite is called *measurements methods*. The reason why it is called *diagnostics* in plasma physics is the following. The same signal observed from a plasma (such as, for example, a spectrum of electromagnetic radiation) could correspond to many different sets of parameters of the plasma and/or of the fields in it. This is analogous to medicine, where a set of patient symptoms could correspond to different maladies. Because of this analogy, that is, because of the nonexistence of one-to-one correspondence between the signals/symptoms and the underlying cause in both plasma physics and medicine, the medical term *diagnostics* is used in plasma physics.

Diagnostic methods using the Stark effect play a prominent role in plasma research—both in laboratories and in astrophysics. They are based on the analysis of the *shape* of spectral lines emitted by plasmas. There are many reasons why the spectral lines emitted by plasmas do not look like a set of line components separated from each other, as in figure 6.1, but rather appear as a continuous profile (or contour), sometimes having a bell-shape (such as in figure 6.2) or sometimes more complicated forms (such as the experimental profile in figure 6.3).

One of the reasons why the spectral lines emitted by plasmas appear as continuous profiles is that the emission is collected from a large number of radiating atoms or ions (hereafter called radiators) along the line of sight of the analyzing device (spectrometer). Even in the simplest (although unrealistic) case where the electric field in a plasma is due only to plasma ions (that is, disregarding the electric field of plasma electrons), each radiator generally experiences a different ion electric field (in terms of magnitude and direction) compared to the ion electric field experienced by other radiators. Therefore the total emission of a particular spectral

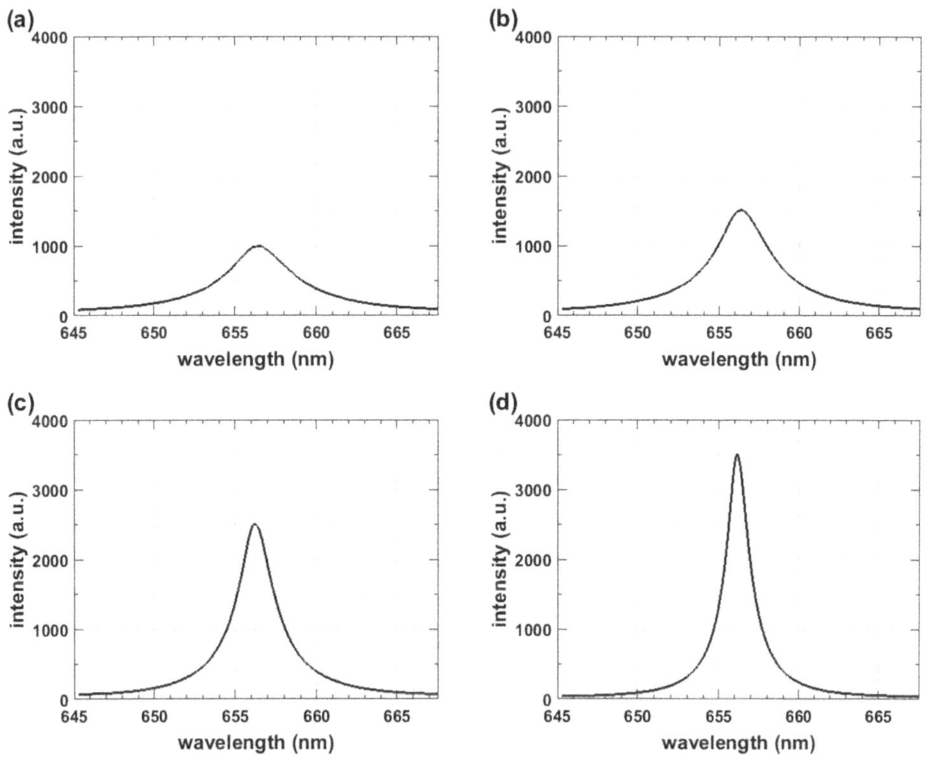

Figure 6.2. Experimental profiles of the hydrogen Balmer-alpha line (H_α) obtained after laser-induced optical breakdown (an intense laser was focused in a cell filled with hydrogen gas at 75% of atmospheric pressure). Profiles (a), (b), (c), and (d) were recorded at time delays (after the breakdown) of 50, 75, 150, and 275 ns, respectively. (Reproduced with permission from [4]. Copyright 2017 C G Parigger, C M Helstern, K M Drake, and G Gautam.)

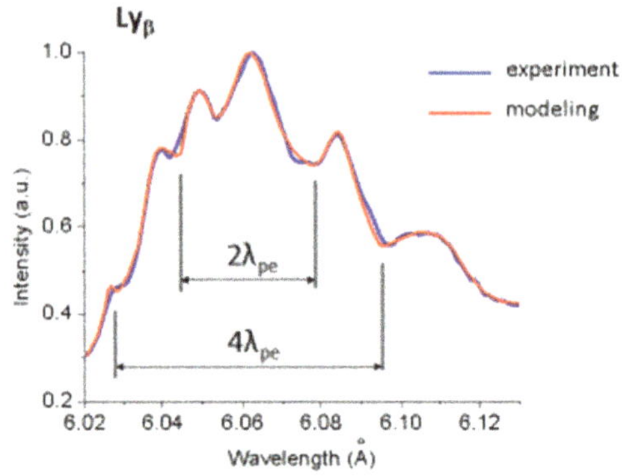

Figure 6.3. Experimental profile (blue line) of the Lyman-beta line, resulting from the quantum radiative transition between the energy levels $n = 3$ and $n = 1$ of the hydrogen-like ion of aluminum (the latter is aluminum that has lost 12 of its 13 electrons)—this spectral line is in the x-ray range. The vertical lines mark features caused by joint action of the high-frequency and low-frequency waves in the plasma. The profile was recorded as a high-intensity laser irradiated aluminum foil. The laser delivered about 10^{21} W (that is, sextillion watts) of power per square centimeter of the target. Such laser intensity not only produced a plasma from the aluminum foil, but also caused plasma electrons to have relativistic velocities, which is why such lasers are called relativistic. The red line shows the corresponding theoretical profile. (Reproduced with permission from [5]. Copyright 2017.)

line would consists of numerous Stark splitting patterns, each one as in figure 6.1, but with different separations between the line components and with different statistical weights. All these numerous Stark splitting patterns would superimpose (that is, lie over one another) such that the individual line components would not show up in the sum. This outcome is called the *Stark broadening* of the spectral line, rather than the Stark splitting.

Moreover, the reality is more complicated than this simple case. The electric fields in plasmas are not only due to relatively slow moving ions, but are also due to relatively fast moving electrons. In addition, there are often also electric fields due to the various waves of different frequency ranges in plasmas. And last, but not least, there could also be the electric field of laser radiation penetrating into plasmas from the outside.

This entire 'zoo' of various types of electric fields in plasmas has to be diagnosed. They can be diagnosed through various types of Stark effects/broadening of spectral lines (as presented, for example in my book *Diagnostics of Laboratory and Astrophysical Plasmas Using Spectral Lineshapes of One-, Two-, and Three-Electron Systems* [6]).

The plasma diagnostic methods based on various types of Stark effects/broadening of spectral lines play a very important role in plasma research for at least four reasons (as presented, in more detail, in [6]). First, these methods are *nonintrusive* (sometimes called *nonperturbing*) because they do not affect the parameters of a plasma (and/or of the fields in it) to be measured—in distinction to the overwhelming majority of other diagnostic methods. Second, these methods are often the *most informative*. Third, they do not depend on model assumptions about the plasma state—in contrast to the diagnostic methods using various ratios of intensities (such as, for example, ratios of integrated intensities of several spectral lines or line-to-continuum ratios). Fourth, sometimes there is *no other method* for the experimental determinations of a particular parameter of the plasma and/or of a particular field in the plasma.

I hope I have convinced the reader of the high practical importance of the Stark effect. Now let us go back to the simplest case of the Stark splitting of hydrogen spectral lines in a static electric field, and try to understand it in relatively simple terms. The phrases 'quantum results' and 'simple terms' are usually as incompatible as the words 'airline' and 'food' in 'airline food' or 'resident' and 'alien' in 'resident alien'. Most often the best way to understand a quantum phenomenon is through its classical analogue (if the latter is possible).

So, let us discuss the classical (rather than the quantum) Stark effect. It should be emphasized that the classical description is quite relevant to the Stark effect in Rydberg atoms, that is, in highly excited atoms ($n \gg 1$).

At the absence of an applied electric field, the electron in a Rydberg atom revolves around the nucleus (the proton) on an elliptical orbit (such as the one shown in figure 3.1), the nucleus being in the focus of the ellipse, see figure 6.4. The unperturbed system (that is, the system in the absence of perturbation, such as an external electric field) has two conserved vector quantities: the angular momentum vector **M** and the Runge–Lenz vector **A**—see figure 3.1 and the paragraph above figure 3.4.

Strictly speaking, the electron in a Rydberg atom would go around the same elliptical orbit over and over again only if we were to disregard relativistic corrections. With the allowance for the relativistic corrections, the electron orbit becomes the one shown in figure 3.5. Namely, the elliptical orbit *slowly* rotates in the plane of the orbit; the ellipse *precesses*. The frequency of the precession is much smaller than the frequency of the revolution of the electron around the ellipse.

A similar, although not identical situation occurs when a Rydberg atom is under a relatively small static electric field **F** (even if one disregards the relativistic corrections). Only the projections of the angular momentum vector **M** and of the Runge–Lenz vector **A** on the electric field vector **F** remain conserved. The projections of the angular momentum vector **M** and of the Runge–Lenz vector **A** on the plane perpendicular to vector **F** do not conserve; these projections precess with the frequency linearly proportional to **F**, mathematical details are presented in

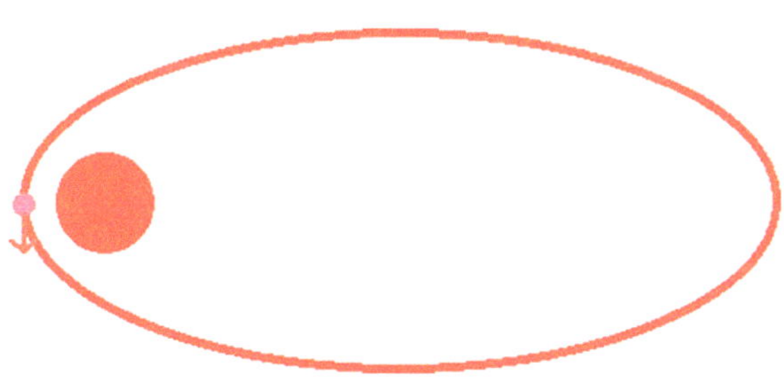

Figure 6.4. Elliptical orbit of the electron in a Rydberg atom around the nucleus at the focus of the ellipse or of a planet around a star, in the absence of perturbation.

appendix C. As a result, the elliptical orbit of the electron also undergoes some kind of precession, the frequency of its precession is the same as for the components of vectors **M** and **A** perpendicular to **F**. In appendix C it is shown that the *classical* precession frequency is

$$\Omega_{\text{class}} = 3e\mathbf{F}/[2(2m_e|E|)^{1/2}], \tag{6.1}$$

where $|E|$ is the absolute value of the energy of the system.

If one were to take the quantum equation (4.1) for the energy and substitute it in equation (6.1), the result would be

$$\Omega_{\text{quant}} = 3n\hbar\mathbf{F}/(2m_e e). \tag{6.2}$$

Amazingly it turns out that the *quantum* frequency Ω_{quant}, obtained in this way from the *classical* precession frequency Ω_{quant}, coincides *exactly* with the separation between Stark sublevels of hydrogen atoms in a relatively small electric field **F**.

Figure 6.5 shows the classical orbit of the electron in a Rydberg atom subjected to a relatively small electric field **F** for a particular case where the field **F** is perpendicular to the angular momentum vector **M**. In this case the electron orbit remains in the plane perpendicular to **M**. The direction of the vector **M** remains unchanged, apart from the sign (although the magnitude does not remain constant). From figure 6.5 it is seen that the electron orbit is an ellipse oscillating (that is, going back and forth) about the direction of the electric field **F** (in figure 6.5, **F** points up along the vertical line).

The oscillations (or more exactly, swings) of the ellipse are confined between the two inclined straight lines. As the major axis of the ellipse reaches one of these straight lines, the direction of the motion of the electron along the ellipse, as well as

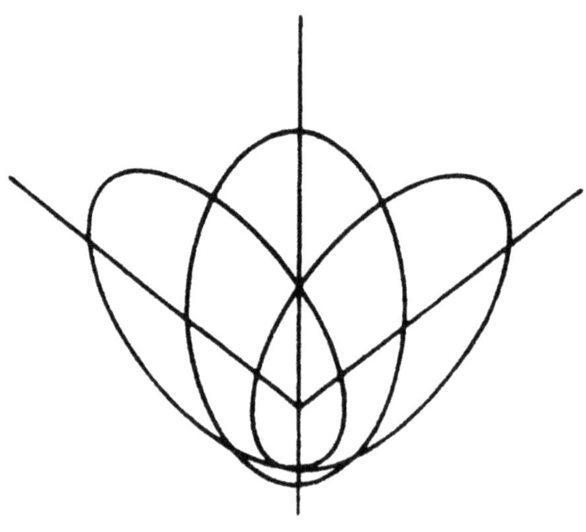

Figure 6.5. Oscillations of the elliptical electron orbit in a Rydberg atom subjected to a relatively small electric field **F** for a particular case where the field **F** is perpendicular to the angular momentum vector **M**. The electric field **F** points up along the vertical line.

the sign of the angular momentum, change to the opposite one and the ellipse swings toward the other straight line.

In the general case of an arbitrary orientation of the applied electric field **F** with respect to the angular momentum **M** of the unperturbed motion, the electron orbit is significantly more complicated. The direction of the angular momentum vector **M** does not remain unchanged and the electron orbit does not remain in-plane. The electron motion is best described not in the Cartesian coordinates (x, y, z), but in the so-called parabolic coordinates. So, let us first briefly discuss different coordinate systems.

Figure 6.6 shows three mutually perpendicular axes: x, y, and z. These are the axes of the Cartesian coordinate system. The location of any point P in three-dimensional space can be completely described by the projections of the radius-vector r (directed from the origin to point P) on these three axes.

There is also a spherical polar coordinate system. In this coordinate system, the location of point P is completely described by its distance r from the origin, by the angle θ that vector **r** makes with the z-axis, and by the angle ϕ that the projection of vector **r** on the xy-plane makes with the x-axis, as shown in figure 6.6.

In the parabolic coordinate system, the location of point P is completely described by the parabolic coordinates (ξ, η, ϕ), where the angle ϕ is the same as in the spherical polar coordinate system and

$$\xi = r + z, \qquad \eta = r - z. \tag{6.3}$$

The surfaces of the constant values of ξ and η are paraboloids of revolution about the z-axis, as shown in figure 6.7.

If the z-axis is chosen along the electric field **F** and the electric field is relatively small, then in the plane containing vector **F** the electron motion is confined, on the one hand, between the some minimum (ξ_{min}) and maximum (ξ_{max}) values of the parabolic coordinate ξ, and on the other hand, between the some minimum (η_{min}) and maximum (η_{max}) values of the parabolic coordinate η, as shown in figure 6.8. Within this plane, the electron orbit is a swinging ellipse.

The corresponding three-dimensional electron orbit can be visualized as follows. Imagine that the curvilinear rectangle (or more rigorously, the curvilinear

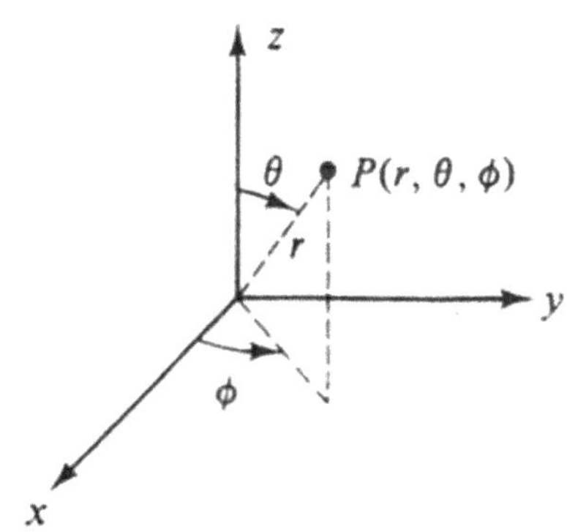

Figure 6.6. The Cartesian coordinate system (x, y, z) and the spherical polar coordinate system (r, θ, ϕ).

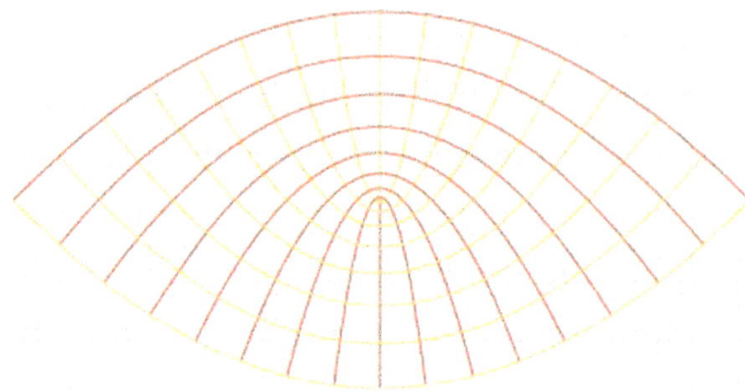

Figure 6.7. The cross-section of the paraboloidal surfaces of constant values of the parabolic coordinates ξ (shown in purple) and η (shown in yellow) by the plane containing the z-axis, which is vertical in this figure.

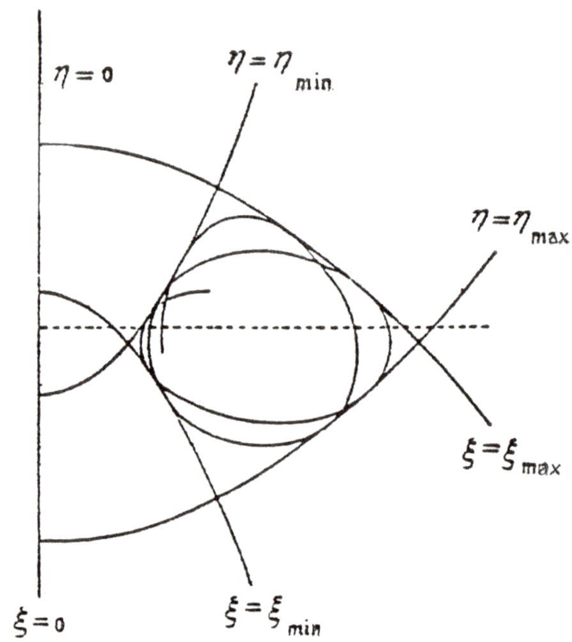

Figure 6.8. Cross-section of the electron orbit in a hydrogen Rydberg atom by the plane containing the electric field **F** (the latter pointing vertically up) for an arbitrary orientation of the vector **F** with respect to the angular momentum **M** of the unperturbed motion.

parallelogram) confining the electron orbit in this plane rotates—together with the electron orbit—about the vertical direction of the electric field **F**. In this way, the electron orbit fills a three-dimensional geometric figure called a torus, which is similar to a doughnut, but with a weird shape (a curvilinear parallelogram) in cross-section. The electron orbit is a swinging ellipse in the cross-sectional plane that simultaneously rotates around the torus/doughnut.

It is important to emphasize the following. Both the swings of the ellipse in the vertical plane and the rotation of the ellipse around the torus occur with the same frequency Ω_{class} (presented in equation (6.1)), as in the case of the much simpler motion depicted in figure 6.5, that is, in the case where the field **F** is perpendicular to the angular momentum vector **M** and the electron orbit is two-dimensional, being confined in the plane perpendicular to vector **M**.

The most amazing thing is that the orbit of a satellite revolving around a planet with a relatively small additional constant acceleration **a**, has the *same possible shapes* as the electron orbit in a hydrogen Rydberg atom in a relatively small electric field **F**. Namely, the satellite orbit is a swinging ellipse in the cross-sectional plane (the plane containing vector **a**) that simultaneously rotates around the torus. Both the swings of the ellipse in the vertical plane and the rotation of the ellipse around the torus occur with the same frequency $\Omega_{satellite}$, where

$$\Omega_{satellite} = (3\mathbf{a}/2)[m/(2|E|)]^{1/2}. \quad (6.4)$$

Equation (6.4) is easily obtained from equation (6.1) by substituting the electron mass m_e with the satellite mass m and by substituting the product $e\mathbf{F}$ with the product $m\mathbf{a}$.

There is, however, an important difference between the electron motion in hydrogen Rydberg atoms subjected to a static electric field **F** and a satellite revolving around a planet with an additional constant acceleration **a**. As long as the acceleration **a** is relatively small, such that the frequency $\Omega_{satellite}$ is much smaller than the frequency of the revolution of the satellite around the planet, the satellite would remain around the planet practically forever (if we were to disregard the effect of other celestial bodies on the satellite–planet system under consideration).

In distinction, for the electron in the Rydberg atom, regardless of how small the electric field **F** is (still satisfying the condition that the frequency Ω_{class} from equation (6.1) is much smaller than the frequency of the revolution of the electron around the nucleus), there is always a chance to escape from the nucleus and become free—an atom will become an ion. This is possible due to the *quantum effect of tunneling* discussed in chapter 3. Here follow some details on tunneling in the case of the Stark effect in Rydberg atoms.

The projection of the three-dimensional potential energy surface on the z-axis (chosen along **F**) can be represented in the form

$$U(z) = U_0(z) + V(z), \quad U_0(z) = -A/|z|, \quad V(z) = Bz, \quad (6.5)$$

where

$$A = e^2, \quad B = e\mathbf{F}. \quad (6.6)$$

(In the case of a satellite around a planet, it would be $A = GMm$, $B = m\mathbf{a}$; here G is the gravitational constant and M is the planet mass.) In equation (6.5), $U_0(z)$ is the unperturbed potential and $V(z)$ is the small perturbation.

In figure 6.9, the plot of the unperturbed potential is shown by the dashed line. The thin horizontal line represents a particular energy level of the hydrogen Rydberg atom. It is seen that the absence of the electric field **F**, for the atomic electron to

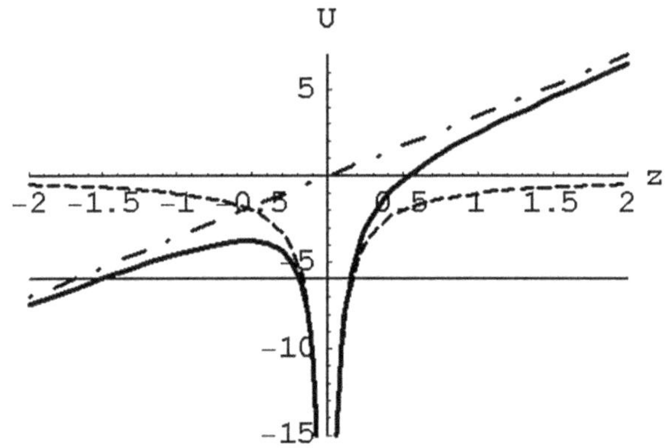

Figure 6.9. The projection $U(z)$ of the three-dimensional potential energy surface on the z-axis chosen along the electric field **F**. The unperturbed potential energy $U_0(z)$ is shown by the dashed line. The perturbation $V(z)$ is shown by the dash-dotted line. The total potential energy $U(z) = U_0(z) + V(z)$ is shown by the solid line. The thin horizontal line represents a particular energy level of the hydrogen Rydberg atom.

escape to 'freedom' (that is, to a very large distance from the nucleus), it would have to go through a barrier of the infinite width. According to quantum mechanics, the probability of going through a barrier of infinite width is zero.

In figure 6.9, the dashed-dotted line is a plot of the perturbation $V(z)$ and the thick solid line is the plot of the total potential energy $U(z) = U_0(z) + V(z)$. It is seen that the presence of the electric field **F** (regardless of how small the field **F** is) opens up an escape channel for the atomic electron. Indeed, in the direction of the negative values of z, the electron encounters a barrier of a *finite* width. According to quantum mechanics, there is a non-zero probability for a particle to go through a barrier of the finite width (the phenomenon of tunneling, the interpretation of which was given in chapter 3). The larger the electric field **F**, the greater the slope of the dash-dotted line, and the smaller the width of the barrier for a given energy level, so the probability of the escape would be greater (that is, the probability of ionization).

So, why is tunneling practically impossible for the case of a satellite around a planet, despite the potential energy $U(z)$, where the z-axis is along the acceleration vector **a**, having the same form as in equation (6.5) and figure 6.9 being equally applicable to this case? To answer this question let us recall the interpretation of tunneling given in chapter 3. Namely, the uncertainty of the kinetic energy $\Delta K \sim \hbar v/\Delta z$ (see equation (3.6)) should be sufficiently large. The uncertainty ΔK is proportional to the velocity v of the particle and the velocity is inversely proportional to the mass of the particle. Since the electron mass m_e is relatively small, ΔK can be sufficiently large to facilitate the escape (as explained in chapter 3). However, the satellite mass m is very large compared to the electron mass m_e: the ratio m/m_e is an astronomically large number. Therefore, in the case of the satellite, the uncertainty of the kinetic energy ΔK is so miniscule that escape would not occur during the lifetime of the satellite or even during the lifetime of the planet.

References

[1] Beletsky V V 2001 *Essays on the Motion of Celestial Bodies* (Basel: Birkhäuser/Springer)
[2] Bialu S and Loeb A 2018 *Astrophys. J. Lett.* **868** L1
[3] Mark H and Wierl R 1929 *Z. Phys.* **53** 526
Mark H and Wierl R 1929 *Z. Phys.* **55** 156
Mark H and Wierl R 1929 *Z. Phys.* **57** 404
[4] Parigger C G, Helstern C M, Drake K M and Gautam G 2017 *Int. Rev. At. Mol. Phys.* **8** 73
[5] Oks E *et al* 2017 *Opt. Express* **25** 1958
[6] Oks E 2017 *Diagnostics of Laboratory and Astrophysical Plasmas Using Spectral Lineshapes of One-, Two-, and Three-Electron Systems* (Singapore: World Scientific)

Eugene Oks

Chapter 7

Similarities between the motion of a satellite around the oblate Earth and Rydberg atoms in a high-frequency laser field and heavy hydrogen-like ions, including a description using uncommon 'two-dimensional' numbers

The Earth is not a sphere. Its equatorial diameter of 12 756 km is 42 km larger than its polar diameter of 12 714 km. The shape of the Earth resembles a three-dimensional object called an *oblate spheroid*, see figure 7.1.

An oblate spheroid can be thought of as an ellipse (such the one in figure 7.2) rotated about its minor axis.

The slightly flattened shape of the Earth is a combined effect of gravitational and centrifugal forces. A few words about centrifugal force might be in order.

If you are in the passenger seat of a car and the driver makes a left turn (or a right turn in the United Kingdom) and you are not wearing the seat belt, then you would feel like some force is throwing you toward the car door next to you. This is an example of centrifugal force. However, from the point of view of an immobile observer in the street, you just continued moving by inertia along the same line as before the driver started turning the car and then the car door simply crossed your straight path as a result of the car turning. In other words, from the point of view of this observer, there was no centrifugal force—you continued moving by inertia. The point is that the centrifugal force is actually a fictitious force; it does not exist in a reference frame that is at rest or moves with a constant (in terms of magnitude and direction) velocity. It exists only in non-inertial reference frames, such as a turning car or the rotating Earth.

Now, back to the motion of the satellite. Because of this slightly flattened shape, the potential energy $U(r)$ of a satellite in the gravitational field of the Earth differs slightly

doi:10.1088/2053-2563/aafed9ch7

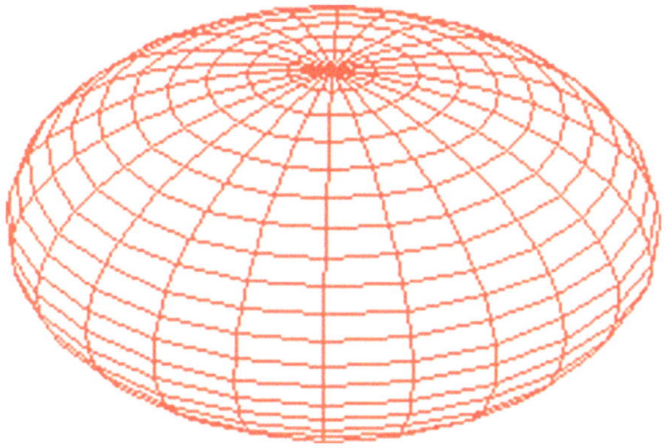

Figure 7.1. Oblate spheroid.

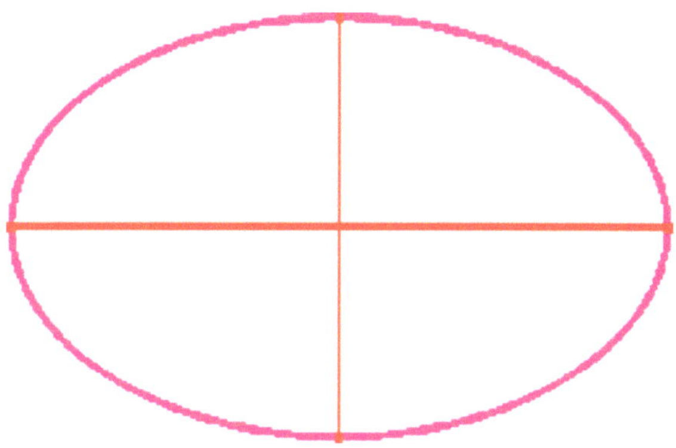

Figure 7.2. Ellipse and its major (horizontal in the figure) and minor (vertical in the figure) axes.

from the potential energy $U_0(r) = GmM/r$ it would have if the Earth were a sphere. The perturbed potential energy $U(r)$ can be *approximately* represented in the form

$$U(r, \phi) = -(GmM/r)[1 - I_2(R/r)^2 P_2(\sin\phi)]. \tag{7.1}$$

(Equation (7.1) is approximate because it represents only the first two terms of the expansion of the potential energy in inverse powers of r.) Here $G = 6.6726 \times 10^{-8}$ $\text{cm}^3\ \text{s}^{-2}\ \text{g}^{-1}$ is the gravitational constant (one of the fundamental constants of nature); m and M are the masses of the satellite and the Earth, respectively; $I_2 = -1.082 \times 10^{-3}$ is a constant related to the slight difference between the equatorial and polar diameters of the Earth; $R = 6.378 \times 10^8$ cm is the equatorial radius of the Earth; θ is the geographical latitude of the satellite at any point of its orbit; and $P_2(\sin\phi)$ is one of the Legendre polynomials:

$$P_2(\sin\phi)] = (3\sin^2\phi - 1)/2. \tag{7.2}$$

For the potential energy from equation (7.1) there exists an exact analytical solution for the satellite's motion. Details can be found, for example, in section 1.7

of Beletsky's book *Essays on the Motion of Celestial Bodies* [1]. The outcome is as follows.

The elliptical orbit of the satellite undergoes two types of precession simultaneously, but *without changing its shape*. The first one is the precession of the orbit in its plane—pretty much as depicted in figure 3.5. It occurs with the angular frequency

$$\Omega_{\text{precession in plane}} = (3I_2/4)(R/p)^2(1 - 5\cos^2 i), \tag{7.3}$$

where i is the *inclination*, that is, the angle between the plane of the satellite orbit and the equatorial plane of the Earth. The quantity p in equation (7.3) is related to parameters of the unperturbed elliptical orbit of the satellite as follows

$$p = 2\, r_{\min}\, r_{\max}/(r_{\min} + r_{\max}), \tag{7.4}$$

where $r_{\min}$ and $r_{\max}$ are, respectively, the minimum and maximum distances of the satellite from the center of the Earth.

Interestingly enough, for $i = 63.4°$, one obtains $1 - 5\cos^2 i = 0$, so that $\Omega_{\text{precession in plane}}$ vanishes. This means that satellites launched at this inclination do not undergo precession in the plane of the orbit.

The second simultaneous precession is the precession of the plane of the satellite orbit, as illustrated in figure 7.3. The precession frequency is

$$\Omega_{\text{precession of plane}} = (3I_2/2)(R/p)^2 \cos i. \tag{7.5}$$

At $i = 90°$ one obtains $\Omega_{\text{precession of plane}} = 0$. For such an inclination, which corresponds to the plane of orbit perpendicular to the Earth equator, there is no precession of the plane of orbit. These are so-called polar satellites.

In summary, the fact that the Earth is slightly flattened does not affect the elliptical shape of the satellite orbit or the inclination of the orbit. It does result, generally, in two simultaneous precessions of the orbit with frequencies given by equations (7.4) and (7.5).

As stated in the title of chapter 7, this celestial problem has two analogies in the micro-world. The first analogy relates to a hydrogen Rydberg atom in the *linearly polarized* electric field of *high-frequency* laser radiation $\mathbf{E}(t) = \mathbf{E}_0 \cos \omega t$ ('linearly polarized' means that the electric field oscillates along a straight line). Here the laser frequency ω is assumed to be much greater than the frequency ω_{Kepler} of the

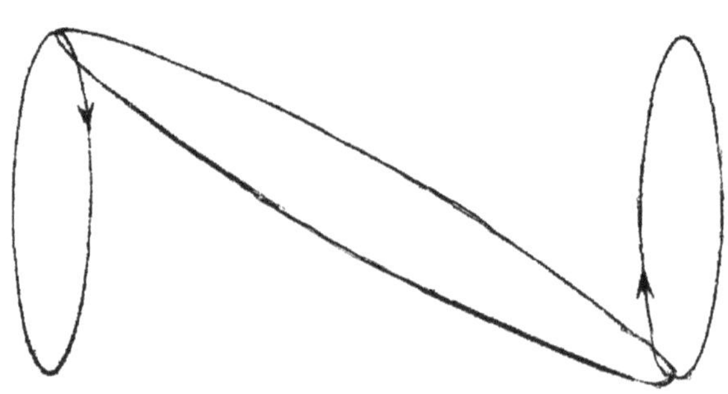

Figure 7.3. Precession of the plane of satellite orbit.

revolution of the atomic electron around the nucleus: $\omega \gg \omega_{\text{Kepler}}$. (The frequency ω_{Kepler} is called the Kepler frequency because it is mathematically equivalent to the frequency of the revolution of a planet around a star and the latter problem was studied by German mathematician and astronomer Johannes Kepler in the 17th century.)

A hydrogen Rydberg atom in a field of high-frequency laser radiation is just one of many examples of a situation where a physical system consists of interacting 'rapid' and 'slow' subsystems, the rapid subsystem being characterized by a much higher frequency than the slow subsystem. For the example under consideration, the rapid subsystem is the laser radiation and the slow subsystem is the hydrogen Rydberg atom.

Physically, the slow subsystem cannot follow the fast time evolution of the rapid subsystem. Instead, the slow subsystem reacts only to some *averaged effect* of the rapid subsystem, the averaging being over intervals of time of the order of the period of the rapid subsystem.

In 1986 Nadezhdin and Oks, in a paper in *Soviet Technical Physics Letters* [2], further developed this averaging method and applied it to a hydrogen Rydberg atom in the linearly polarized electric field of high-frequency laser radiation. (Later Oks, Davis, and Uzer, in a paper published in *Journal of Physics B: Atomic, Molecular, and Optical Physics* [3], applied this averaging method to a hydrogen Rydberg atom in a circularly polarized or elliptically polarized electric field of high-frequency laser radiation—details can be found in appendix D.) For the linearly polarized laser field, Nadezhdin and Oks showed that the perturbed motion of the atomic electron can be approximately characterized by the following potential energy (in spherical polar coordinates with the z-axis along vector $\mathbf{E}_0$):

$$U(r, \theta) = -e^2/r - (\gamma/r^3)(3\cos^2\theta - 1), \quad \gamma = e^4E_0^2/(4m_e^2\omega^4), \tag{7.6}$$

where θ is the polar angle, that is, the angle between the radius-vector $\mathbf{r}$ of the electron and the z-axis.

Taking into account that the geographical latitude ϕ in equation (7.1) and the polar angle θ in equation (7.6) are related as $\phi = \pi/2 - \theta$ (where $\pi/2$ is the 90° angle expressed in radians), so that $\sin^2\phi = \cos^2\theta$, and that the quantity I_2 in equation (7.1) is negative (so that $I_2 = -|I_2|$), equation (7.1) can be re-written in the form

$$U(r, \theta) = -(GmM/r) - (GmM \mid I_2 \mid R^2/r^3)(3\cos^2\theta - 1)/2. \tag{7.7}$$

The comparison of equations (7.6) and (7.7) shows that the potential energy of the electron in a hydrogen Rydberg atom under a high-frequency linearly polarized laser field is mathematically equivalent to the potential energy of a satellite around the oblate Earth. Indeed, if in equation (7.7) one were to substitute GmM with e^2 and $|I_2|R^2$ with $2\gamma/e^2$, one would obtain equation (7.6).

Therefore, the motion of the electron in a hydrogen Rydberg atom under a high-frequency linearly polarized laser field can be described in the same way as the motion of a satellite around the oblate Earth. First, the shape of the elliptical orbit and the angle between the orbital plane and the plane perpendicular to the laser field amplitude $\mathbf{E}_0$ would not be affected by the laser field; they would remain the same over the course of time. The elliptical orbit of the electron undergoes two precessions simultaneously: the elliptical orbit precesses in its own plane with the frequency $\Omega_{\text{precession in plane}}$ (similarly to the depiction in figure 3.5) and the plane of the orbit precesses around the laser field amplitude $\mathbf{E}_0$ with the frequency $\Omega_{\text{precession of plane}}$ (similarly to the depiction in figure 7.3), both frequencies being proportional to $\mathbf{E}_0^2$.

There is an interesting fact about the conserved quantities for both of these physical systems. If the motion of a particle is characterized by a potential energy that depends only on the magnitude r of the radius-vector $\mathbf{r}$ of the particle, but not on the direction of vector $\mathbf{r}$, then all directions in space are equivalent for such a system—the system is said to have *spherical symmetry*. As a consequence of spherical symmetry, the angular momentum vector $\mathbf{M}$ is conserved both by the magnitude and by the direction for such a system. If a linearly polarized laser field is applied to such an atomic system, then there is no longer an equivalence of all directions in space; the potential energy now depends not only on the magnitude r of the radius-vector $\mathbf{r}$, but also on the angle θ between the vector $\mathbf{r}$ and the laser field amplitude vector $\mathbf{E}_0$. The latter vector defines the preferred direction in space. Only all directions in the plane perpendicular to the vector $\mathbf{E}_0$ remain equivalent—the symmetry of the system is said to be reduced from spherical to axial. As a consequence, only the angular momentum projection M_z on vector $\mathbf{E}_0$ remains conserved.

Similarly, because the Earth is not a sphere but an oblate spheroid, the axis connecting the two poles defines the preferred direction in space. Only the angular momentum projection M_z on this direction remains conserved.

However, for the specific form of the potential energy dependence on $\mathbf{r}$ and θ, as in equations (7.6) and (7.7), in addition to the exact conservation the angular momentum projection M_z, there is also the *conservation of the square of angular momentum* M^2. The latter quantity is proportional to the area of the orbit. Therefore, the fact that the shape of the elliptical orbit is not affected by the perturbation (by the laser field for the atomic system or by the spheroidness of the Earth in the case of the satellite) signifies the conservation of M^2.

For systems with physical quantities whose conservation does not follow from the geometrical symmetry of the system (in the above case from the axial symmetry), it is said that such systems have a *higher than geometrical symmetry*. A further discussion of the role of symmetries in physics will be presented in chapter 11.

It should be emphasized that for the type of potential energies given by equation (7.6) or (7.7), the square of the angular momentum M^2 is conserved exactly. However, equations (7.6) and (7.7) were obtained by neglecting some corrections (much smaller than the second term in (7.6) or in (7.7)). Since equations (7.6) and (7.7) are approximate, so is the conservation of M^2. Nevertheless, any additional quantity that is conserved (whether exactly or approximately), is physically important.

Now let us discuss the situation where a hydrogen Rydberg atom is under a *circularly polarized* (rather than linearly polarized) high-frequency laser field. If the

z-axis is chosen to be perpendicular to the polarization plane, then the circularly polarized laser field can be represented in the form

$$\mathbf{E}(t) = \varepsilon_0\left(\mathbf{e}_x \cos \omega t + \mathbf{e}_y \sin \omega t\right), \tag{7.8}$$

where $\mathbf{e}_x$ and $\mathbf{e}_y$ are the unit vectors along the x- and y-axes, respectively. After averaging over the rapid system, the potential energy of the atomic electron takes the following form (as shown by Oks, Davis, and Uzer in [3]):

$$U(r, \theta) = -e^2/r + (\gamma/r^3)(3\cos^2\theta - 1), \quad \gamma = e^4\varepsilon_0{}^2/(8m_e{}^2\omega^4). \tag{7.9}$$

Equation (7.9) differs from equation (7.6) only in the sign of the second term. The 'celestial' analogy is a satellite moving in the potential from equation (7.1), but with the positive constant I_2, which corresponds to the case of a planet in the shape of a *prolate* (rather than an oblate) spheroid, as depicted in figure 7.4. A prolate spheroid can be thought of as an ellipse (such as the one in figure 7.2) rotated about its major axis—in contrast to the oblate spheroid resulting from an ellipse rotated about its minor axis.

It should be emphasized that a rotating planet cannot have the form of a prolate spheroid, but only the form of an oblate spheroid. This is why in the previous paragraph the word 'celestial' was in quotation marks.

However, the nuclei of heavy atoms can have either the shape of a prolate spheroid or the shape of an oblate spheroid. So, the motion of the electron in a hydrogen Rydberg atom under a high-frequency laser field with linear or circular polarization also has an analogy with the motion of the electron in heavy hydrogenic ions.

It could be noted in passing that there are hydrogenic atoms and ions where the negatively charged particle orbiting the nucleus is not the electron, but the muon. The muon is a 'brother' of the electron, being about 200 times heavier than the

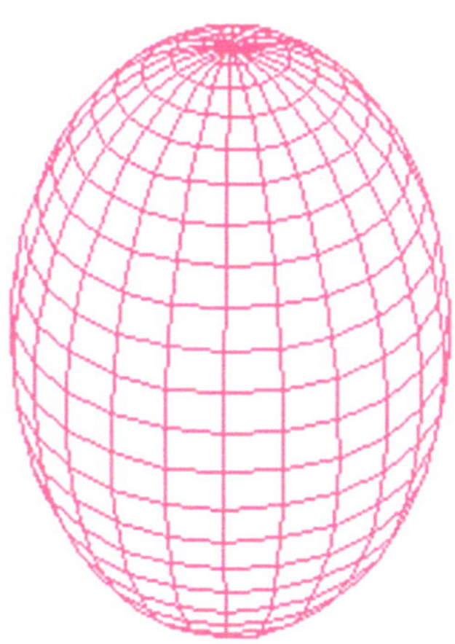

Figure 7.4. Prolate spheroid.

electron. Muonic atoms and ions are more sensitive to the nuclear shape than electronic atoms and ions. In particular, in a paper published in *Atoms* [4], the current author showed how the shift of spectral lines of muonic hydrogenic ions can serve in the experimental determination of the parameters of nuclear shape (such as, for example, a parameter analogous to the constant I_2 in equation (7.1)).

To conclude this chapter, let us come back to the motion of a satellite around the oblate Earth and look into this problem from a totally unexpected point of view. It turns out that this problem is *in some sense* equivalent to the problem of a particle in the gravitational field of two fixed 'Newton centers', that is, in the field of two massive bodies separated by a distance R, each attracting the particle in accordance to Newton's gravitational law. This fundamental problem was first posed by the Swiss mathematician, physicist, and astronomer Leonhard Euler in the 18th century. In the 19th century, German mathematician Carl Gustav Jacob Jacobi showed that the problem allows an exact analytical solution by the proper choice of coordinates —namely, in the so-called elliptical coordinates. Numerous studies and refinements of the solution have been performed since then and the solution is included in many textbooks on classical mechanics.

The most mind-boggling fact about the analogy between these two physical systems is the following. For the problem of a particle in the field of two fixed Newton centers to become mathematically equivalent to the problem of the motion of a satellite around an oblate planet, the two Newton centers should be considered to be separated by an *imaginary distance*! Before providing a brief explanation, the following joke (or piece of wisdom) might be in order:

> Physicists have a knack for jumping into mathematical cesspools and coming out smelling like a rose.

In addition to *real* numbers, mathematicians also consider so-called *complex* numbers. They were introduced to simplify the form of the solutions of many problems. Complex numbers are two-dimensional; they are represented by a pair of real numbers, such as, for example, (3, 5) or (2.4, −3.9) and so on. Generally, a complex number is denoted as $z = a + ib$, where a and b are real numbers and 'i' is called the imaginary unit. Complex numbers can be better understood by their pictorial representation, as given in figure 7.5.

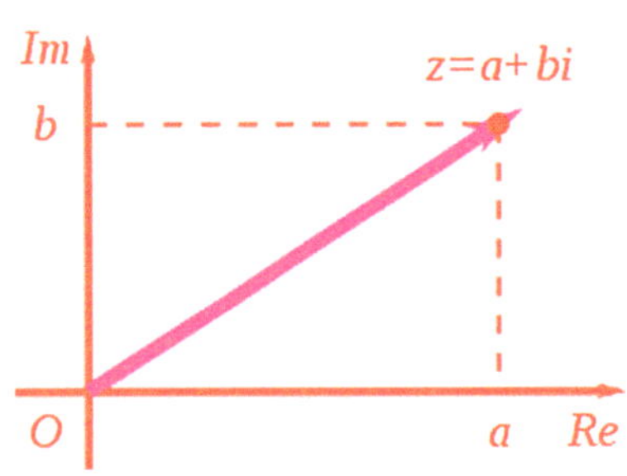

Figure 7.5. Pictorial representation of a complex number $z = a + ib$.

In the pictorial representation, a complex number $z = a + ib$ is a two-dimensional vector. Its projection a on the horizontal axis is called the real part of the complex number, while its projection b on the vertical axis is called the imaginary part of the complex number. From the pictorial representation it is clear that the imaginary unit 'i' has the meaning of the unit vector along the vertical axis.

Complex numbers with a zero real part ($a = 0$) are called purely imaginary or just imaginary numbers. So, when the distance R between the two fixed Newton centers is formally considered to be an imaginary number, then the problem of a particle (or a satellite) in the field of the two fixed Newton centers becomes mathematically equivalent to the problem of a satellite around an oblate planet. Moreover, since the former problem has an exact analytical solution, when used for the problem of a satellite around an oblate planet, it provides a *more accurate solution* for the latter problem than the approximate solution based on the approximate potential energy from equation (7.1).

For completeness we note that the problem of a particle (or a satellite) in the field of two fixed Newton centers, separated by a *real* distance R, is mathematically equivalent to the problem of a satellite around a *prolate* planet (if such a planet were to exist).

References

[1] Beletsky V V 2001 *Essays on the Motion of Celestial Bodies* (Basel: Birkhäuser/Springer)
[2] Nadezhdin B B and Oks E 1986 *Sov. Tech. Phys. Lett.* **12** 512
[3] Oks E, Davis J and Uzer T 2000 *J. Phys. B: At. Mol. Opt. Phys.* **33** 207
[4] Oks E 2018 *Atoms* **6** 14

Chapter 8

The ‘Sun’ going back and forth: similarities between a planet around binary stars or a star–planet–moon system and one-electron Rydberg quasimolecules

We are lucky to live on a planet revolving around a single star. In the Universe there are planets revolving around a *binary star*, which is a system of two stars orbiting around their common center of mass, called the barycenter. On the one hand, observations have shown that the majority of single, Sun-like stars have planets, while only about one third of binary stars have them. On the other hand, it was estimated that about half of binary stars are capable of supporting habitable Earth-like planets, which is why studies of planets in binary star systems are particularly important.

The motion of a planet around a binary star is much more complicated compared to the motion of a planet around a single star. It represents an example of the classical three-body problem: three masses interacting with one another in accordance with Newton’s gravitational law. As early as the 17th century, Newton unsuccessfully tried to solve this problem. In 1887 and 1889, German mathematician and astronomer Ernst Heinrich Bruns and French mathematician and physicist Jules Henri Poincaré showed that there is no general analytical solution to the three-body problem in terms of algebraic expressions and integrals. To make things worse, the motion of three bodies generally does not repeat itself (with few specific exceptions).

Galileo Galilei wrote: ‘Mathematics is the language in which God has written the Universe’. Well, it turns out that already for the three-body problem (and more so for the N-body problem with $N > 3$), the mathematics becomes far more complicated compared to the two-body problem, such as a planet revolving around a single star.

doi:10.1088/2053-2563/aafed9ch8

This is why we are lucky—because of the relative simplicity of the motion of the Earth around the Sun, Kepler and Newton succeeded in figuring out the simple law of gravitation. As Beletsky wrote in *Essays on the Motion of Celestial Bodies* [1], a civilization residing on a planet revolving around a binary star would be in a worse situation—their Keplers and Newtons would observe a very complicated motion, from which it would be much more difficult to come up with the simple gravitational force law $F = GM_1M_2/r_{12}^2$, where M_1 and M_2 are the masses of any two bodies and r_{12} is their distance from each other. This could have slowed down the progress of such a civilization.

Speaking of the millions and millions of various types of stars (single, binary, and so on), reminds me of the following joke:

> Sherlock Holmes and Dr Watson went on a camping trip. After a good meal and a bottle of wine, they lay down for the night and went to sleep. A few hours later, Holmes awoke and nudged his friend.
> 'Watson, look up and tell me what you see.'
> Watson replied: 'I see millions and millions of stars.'
> 'What does that tell you?'
> Watson thought for a minute. 'Astronomically, it tells me that there are millions of galaxies and possibly billions of planets. Astrologically, I observe that Saturn is in the Leo constellation. Horologically, I deduce that the time is approximately a quarter past three. Theologically I can see that God is all powerful and that we are small and insignificant. Meteorologically, I suspect that we will have a beautiful day tomorrow. Why, what does it tell YOU?'
> Holmes was silent for a minute, then said: 'Watson, you idiot! It tells me that some jerk has stolen our tent!'

Coming back to the three-body problem, since the general analytical solution was proven to be impossible, a series of simplifications (or 'particular cases') were considered over the course of time. The first simplification, called the *restricted three-body problem* (introduced by Leonhard Euler), was to consider one of the bodies to have a much smaller ('negligible') mass compared to the two other bodies, so that the third body does not affect the motion of the two other bodies. The restricted three-body problem approximation can be reasonably applied to a planet around a binary star (the planet being the third body).

Even the restricted three-body problem turned out to be too complicated for general analytical treatment. Therefore, the overwhelming majority of works on planetary orbits around binary stars considered only a further simplified situation, called the *planar restricted three-body problem*, where all the three bodies move in the same plane. In other words, the two stars orbit one another and the planet moves within the orbital plane of the stars, the motion of each of the three bodies being essentially *two-dimensional*. Moreover, a significant number of works even used the additional simplifying assumption that the orbits of the stars are circular (despite them generally being elliptical); this simplest situation is called the *circular planar restricted three-body problem*.

In 2015 the author of this book published a paper in *Astrophysical Journal* [2] presenting his discovery of the approximate analytical solution for the *three-dimensional, non-planar restricted three-body problem*. The resulting planetary trajectory is a helix on the surface of a cone, with the axis coinciding with the interstellar axis. In this *conic-helical* state, the planet, while spiraling on the surface of the cone, oscillates between two end-circles which result from cutting the cone with two parallel planes perpendicular to its axis, see figure 8.1. As the two stars revolve about their center of mass, the interstellar axis rotates relatively slowly (slowly compared to the fast motion of the planet) and the 'corkscrew' orbit of the planet follows the rotation of the interstellar axis for a large number of orbital periods of the stars.

The crucial question was how stable this conic-helical orbit is under the slow rotation of the interstellar axis. While a detailed answer will be provided in the following paragraphs, let us emphasize right now that this stability requires that the entire 'corkscrew' planetary orbit should be much closer to the star of smaller mass μ and that the ratio $b = \mu'/\mu$ of the stellar masses should be greater than or of the order of 10. (According to astronomical observations, the actual range of the ratio $b = \mu'/\mu$ of stellar masses in binary stars is up to 100.)

Any possible intelligent life on such a planet would have a very different experience compared to us. First, the change of seasons would be much faster. Second, as the spiraling planet reached one end of the corkscrew orbit and started spiraling toward the other end, the inhabitants would observe the closest rising 'Sun'

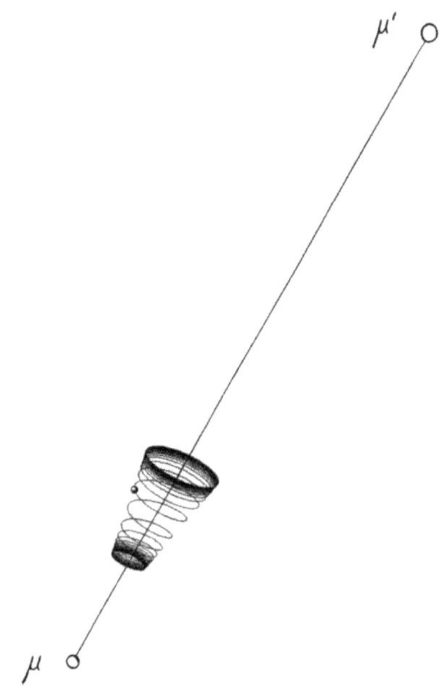

Figure 8.1. Sketch of the conic-helical motion of the planet in the model binary star system, where the star masses are denoted as μ and $\mu' > \mu$. The trajectory has been purposely stretched along the interstellar axis to make its details better visible. (Reproduced with permission from [3]. Copyright 2015 N Kryukov and E Oks.)

suddenly moving backwards. As the planet would be spiraling back and forth between the two ends of the corkscrew, these aliens would see the closest 'Sun' going back and forth in the sky.

For more details on the stability of such planetary system, first of all it is required that the frequency Ω of the rotation of the planet in the plane perpendicular to the interstellar axis should be much greater than the frequency ω of the stars' rotation about their center of mass: $\Omega \gg \omega$. Second, it is required that the coordinate of the planet along the interstellar axis undergoes only relatively small oscillations with respect to the equilibrium position of this coordinate. Similarly, it is required that the radius of the planetary orbit (in the plane perpendicular to the interstellar axis) would undergo only relatively small oscillations with respect to the equilibrium value of the radius.

Corresponding illustrations were provided in a subsequent paper by Kryukov and Oks in *Journal of Astrophysics and Aerospace Technology* [4]. Let us consider a reference frame rotating with the interstellar axis; we choose the z-axis along the interstellar axis in the direction from mass μ to mass μ'. It is convenient to introduce a scaled dimensionless coordinate $w = z/R$, where R is the interstellar distance.

Figure 8.2 shows the scaled dimensionless amplitude $\delta w_0 = \delta z/R$ of the oscillations of the planetary orbit along the interstellar axis versus the scaled projection $w = z/R$ of the average plane of the planetary orbit on the interstellar axis, for three values of the ratio $b = \mu'/\mu$ of the stellar masses. It is seen that $\delta w_0 \ll w$ (i.e. the amplitude of the oscillations along the interstellar axis is much smaller than the equilibrium value of the coordinate z) for b greater than or of the order of 10.

For the validity condition $\Omega \gg \omega$ of this result to be satisfied with a large margin of 'safety', the average plane of the planetary orbit should be very close to the star of smaller mass (and the closer it is in the range of $w < 0.05$, the smaller is the amplitude δw_0, as seen from figure 8.2). As long as the primary frequency Ω of the planet's revolution about the interstellar axis exceeds by many orders of magnitude the Kepler frequency ω of the stars' rotation, the conic-helical planetary orbit would

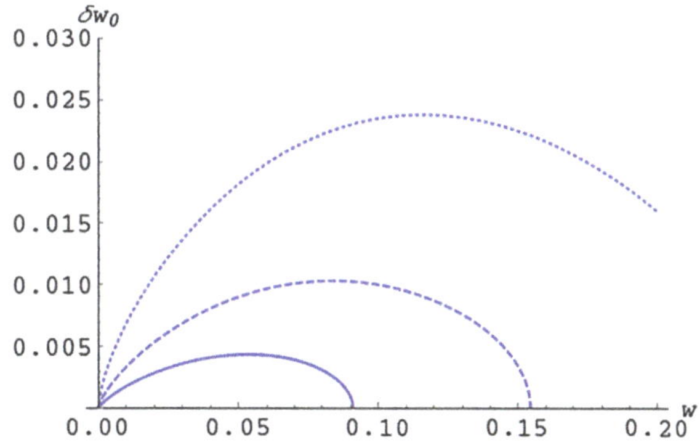

Figure 8.2. The scaled dimensionless amplitude $\delta w_0 = \delta z/R$ of the oscillations of the planetary orbit along the interstellar axis versus the scaled projection $w = z/R$ of the average plane of the planetary orbit on the interstellar axis, for three values of the ratio $b = \mu'/\mu$ of the stellar masses $b = 100$ (solid line), $b = 30$ (dashed line), and $b = 10$ (dotted line).

remain stable for a very long time. (Rigorously speaking, the planet is in a so-called *metastable* state, which is an analogy with atomic/molecular systems—they have metastable states living by many orders of magnitude longer than other states of the system.)

Figure 8.3 shows the scaled dimensionless amplitude $\delta v = \delta r/R$ of the oscillations of the radius r of the planetary orbit versus the scaled projection $w = z/R$ of the average plane of the planetary orbit on the interstellar axis, for three values of the ratio $b = \mu'/\mu$ of the stellar masses. It is seen that for the range of b greater than or of the order of 10 (required to have $\delta w_0 \ll 1$), we have also $\delta v \ll 1$, so that this configuration would remain stable for a very long time.

How long is 'long'? Well, planets do not exist infinitely anyway—they have a finite lifetime; for example, through a collision with an asteroid they could be split into parts or at least be kicked away from the star. So, a metastable corkscrew orbit could exist for a period comparable to the lifetime of the planet.

Here is an example of an astrophysicists' joke about the lifetime of planets:

Old planets never die: they just lose their attraction.

Any planet orbiting a star outside the solar system is called an *exoplanet*. Exoplanets can be detected photometrically if they are *transitable*, meaning that the projections of the planet and of the adjacent star on the plane of the sky coincide once in a while—thus causing a transient variation in the observed light from the star. It turns out that planets in conic-helical orbits of binary stars are 'highly transitable'—meaning that they are transitable for the overwhelming majority of inclinations of the plane of the stars' orbits.

However, by detecting an exoplanet around a binary star in this way it would be nearly impossible to find out whether the plane executes a corkscrew orbit. Potentially, there could be another detection method that would allow us to

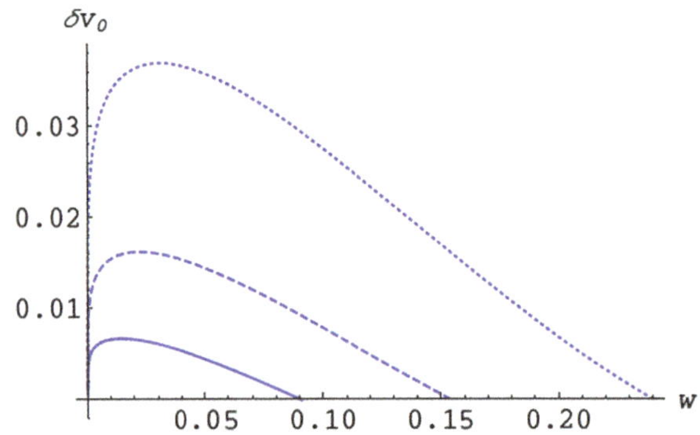

Figure 8.3. The scaled dimensionless amplitude δv_0 of the oscillations of the radius r of the planetary orbit versus the scaled projection $w = z/R$ of the average plane of the planetary orbit on the interstellar axis, for three values of the ratio b of the stellar masses: b = 100 (solid line), b = 30 (dashed line), and b = 10 (dotted line).

recognize a planet on a corkscrew orbit, as follows. In the future, when with progress in gravitational wave detectors it would be possible to detect gravitational waves from binary stars, the presence of a planet in a conic-helical orbit could be revealed by the noticeably *enhanced gravitational radiation* from the binary star system. At first glance, it might seem impossible that the presence of a planet could noticeably enhance the total gravitational radiation from the system—the power P of the gravitational radiation is proportional to the square of the mass, so that due to the small ratio of the masses of the planet and of the stars, the additional gravitational radiation due the presence of the planet might seem to be miniscule. However, the quantity P is also proportional to the sixth power of the orbital frequency, so that the dependence of the power P on the orbital frequency is much stronger than the dependence on the mass. Therefore, since the orbital frequency of the planet in a conic-helical orbit is much greater than the orbital frequency of the stars in the binary, the power P_{planet} of the gravitational radiation due to the planet can be comparable or even exceed the power P_{star} of the gravitational radiation due to the stars.

There are similar celestial systems, such as a moon orbiting a planet revolving around a star. Such systems were also analyzed in Kryukov–Oks paper [4]. They complemented analytical results with exact simulations. The exact simulations demonstrated that the moon can have a practically stable conic-helical orbit around the planet, the average plane of the orbits being perpendicular to the axis connecting the planet and the star. It was a counterintuitive result—under the influence of a relatively distant star, the plane of the moon's orbit could orient itself to be perpendicular to the planet–star axis and the orbit becomes conic-helical. Below are some details of the corresponding simulations.

The time evolution of the moon has been presented in terms of the scaled, dimensionless time

$$\tau = t(G\mu/R^3)^{1/2}, \tag{8.1}$$

where $G = 6.6726 \times 10^{-8}$ cm^3 s^{-2} g^{-1} is the gravitational constant, μ is the planet mass, and R is the planet–star distance. With the z-axis being the planet–star axis, the time evolutions of the two other Cartesian coordinates x and y, scaled by R to become dimensionless quantities $m = x/R$ and $n = y/R$, are presented in figures 8.4 and 8.5 for the initial value of the scaled projection of the average plane of the moon's orbit on the planet–star axis $w_0 = z_0/R = 10^{-4}$. For example, if $R = 1\,500\,000\,000$ km, then $z_0 = 150\,000$ km, which is of the same order of magnitude as the average distance between the Earth and the Earth's Moon.

It is seen that the x- and y-coordinates of the moon orbit undergo relatively simple small oscillations around the planet–star axis. Thus, the motion of the moon is stable in these two directions.

Figure 8.6 shows the dependence of w, the scaled planet–star axial coordinate of the moon (the z-coordinate in the above-mentioned Cartesian coordinate system), on the scaled time τ defined in equation (8.1) for the initial value of the scaled projection of the average plane of the moon's orbit on the planet–star axis $w_0 = z_0/R = 10^{-4}$.

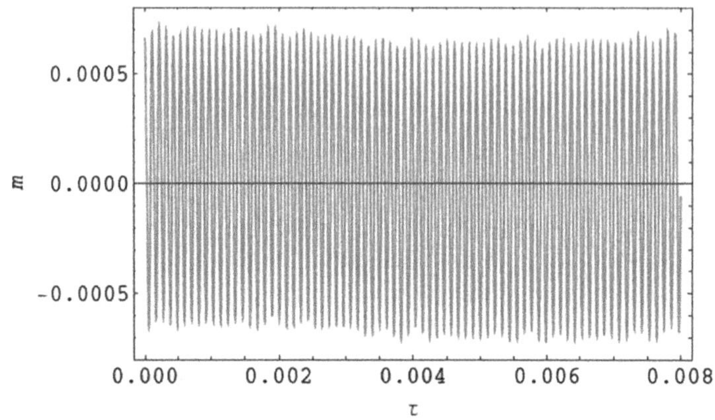

Figure 8.4. Dependence of $m = y/R$, the scaled y-coordinate of the moon, on the scaled time τ defined in equation (8.1) for the initial value of the scaled projection of the average plane of the moon's orbit on the planet–star axis $w_0 = z_0/R = 10^{-4}$.

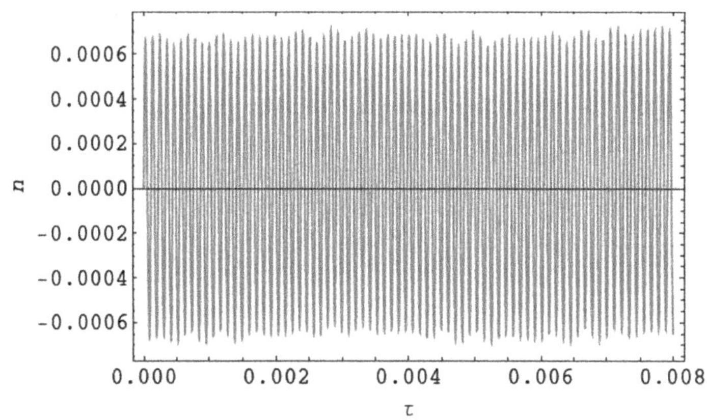

Figure 8.5. Dependence of $n = x/R$, the scaled x-coordinate of the moon, on the scaled time τ defined in equation (8.1) for the initial value of the scaled projection of the average plane of the moon's orbit on the planet–star axis $w_0 = z_0/R = 10^{-4}$.

It is seen that the z-coordinate of the moon orbit (the coordinate along the planet–star axis) undergoes more complicated oscillations than the x- and y-coordinates. However, the maximum amplitude of these oscillations is also limited by a relatively small value—such as the x- and y-coordinates. Thus, the motion of the moon along the planet–star axis is also stable.

Figure 8.7 shows the three-dimensional trajectory of the moon. The planet–star axis is vertical in this figure. The simulated orbit has the conic-helical shape predicted by the analytical results; it is possible to see this in the figure—even though the much smaller size of the z-oscillations of the orbit (compared to the x- and y-oscillations) makes it a little bit difficult (yet feasible) to see.

In summary, it is clearly seen from figures 8.4–8.7 that the conic-helical orbit of the moon in the star–planet–moon system is stable. Thus, under the influence of a

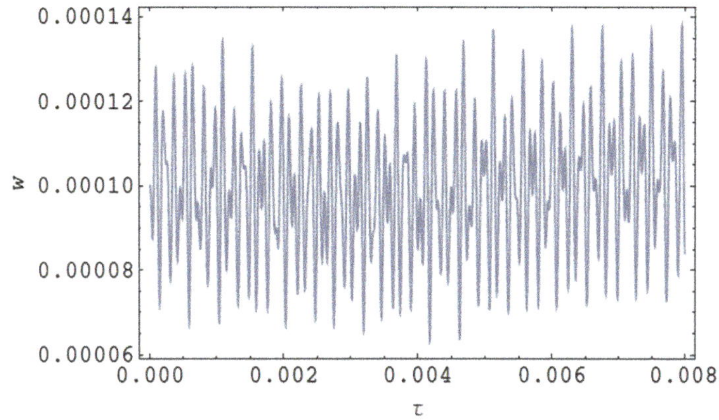

Figure 8.6. Dependence of $w = z/R$, the scaled z-coordinate of the moon, on the scaled time τ defined in equation (8.1) for the initial value of the scaled projection of the average plane of the moon's orbit on the planet–star axis $w_0 = z_0/R = 10^{-4}$.

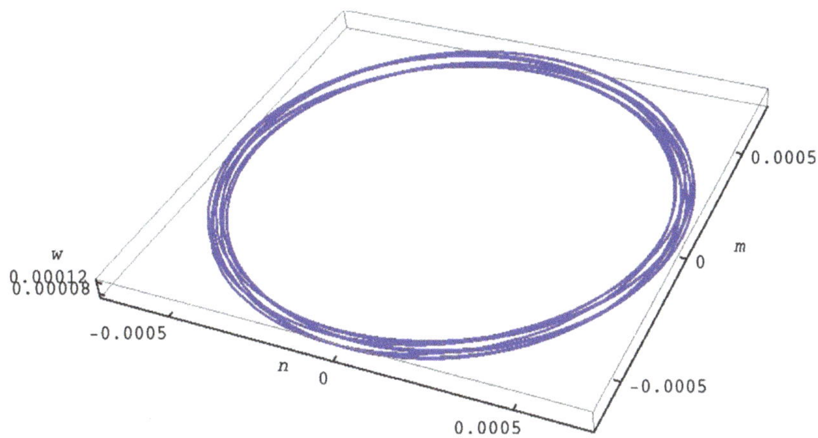

Figure 8.7. Simulated three-dimensional trajectory of the moon in scaled xyz-coordinates (nmw-coordinates) for the initial value of the scaled projection of the average plane of the moon's orbit on the planet–star axis $w_0 = z_0/R = 10^{-4}$. The planet–star axis is vertical in this figure.

relatively *distant star*, the trajectory of the moon becomes conic-helical (a 'cork-screw' shape) and the plane of the moon orbit could orient itself to be *perpendicular* to the planet–star axis. The latter is amazing. Think again about this. The star could be, for example, 10 000 times farther away from the planet than the moon—yet the distant star can dramatically affect the orientation of the moon orbit!

Speaking of planets and moons, it should be noted that moons typically do not have atmospheres while some planets do. There is a relevant joke, as follows:

– Did the astronaut like the restaurant on the Moon?
– He thought the food was fine, but there wasn't much of an atmosphere.

In the same Kryukov–Oks paper [4], they also analyzed whether an object with a corkscrew orbit around a binary star can move with a relativistic velocity (that is, with a velocity comparable to the speed of light c = 300 000 km s^{-1}), and if it can, whether the relativistic corrections (following from Einstein's special relativity theory) would be noticeable. On the basis of the corresponding analytical results obtained, they provided an example of a binary system with star masses $\mu = 1$ and $\mu' = 100$ (in units of the mass of the Sun) and the interstellar distance $R = 100$ astronomical units (a.u.), where 1 a.u. = 15 000 000 km is the average distance between the Earth and the Sun. It should be noted that according to astrophysical observations, it is a realistic example of a binary star system.

According to Kryukov and Oks's analytical results [4], in order for relativistic corrections for corkscrew-type motion of an object around such a binary star to be noticeable, the average radius of the orbit should not exceed 15–30 km. Since the radius of the object should be smaller than the average radius of its orbit, then the object cannot be a planet, but only a so-called *planetoid* of a size not exceeding, say, 10 km.

Kryukov and Oks showed analytically that if the average plane of the planetoid orbit were very close to the lighter star, then such a system would remain stable for a very long time. They also confirmed this conclusion by simulations, some of which are presented below. Figure 8.8 shows the dependence of the scaled y-coordinate $m = y/R$ of the planetoid in the Cartesian xyz-system, with the z-axis being the interstellar axis, on the scaled time τ defined in equation (8.1). The initial value of the scaled projection of the average plane of the planetoid orbit on the interstellar axis was $w_0 = z_0/R = 10^{-25}$, that is, practically coinciding with the position of the lighter star in the binary system.

Figure 8.9 similarly presents the dependence of the scaled x-coordinate $n = x/R$ of the planetoid in the above-mentioned Cartesian xyz-system, on the scaled time τ defined in equation (8.1), for the same initial conditions.

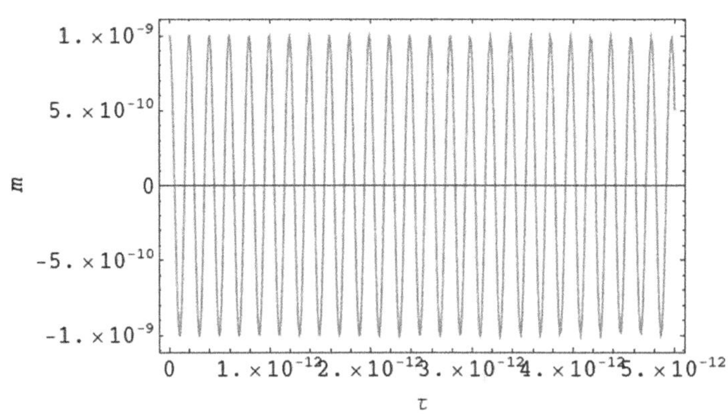

Figure 8.8. Dependence of $m = y/R$, the scaled y-coordinate of the planetoid, on the scaled time τ defined in equation (8.1) for stars of masses $\mu = 1$ and $\mu' = 100$ (in units of the mass of the Sun), the interstellar distance $R = 100$ astronomical units, and the scaled projection of the average plane of the planetoid orbit on the interstellar axis $w_0 = z_0/R = 10^{-25}$.

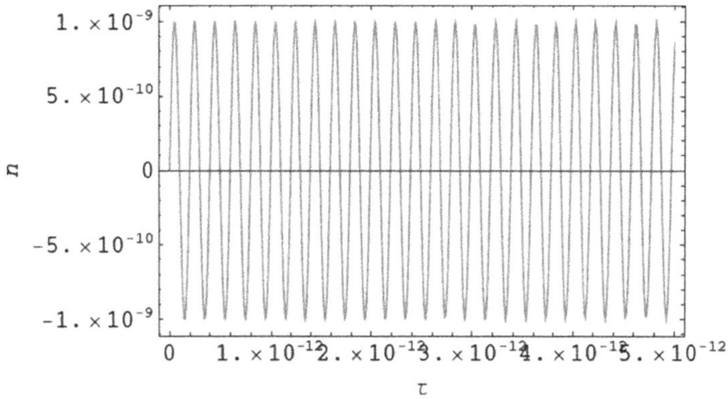

Figure 8.9. The same as in figure 8.8, but for the scaled x-coordinate of the planetoid $n = x/R$.

Figure 8.10 shows the dependence of the scaled z-coordinate of the planetoid $w = z/R$ on scaled time τ defined in equation (8.1), for the same initial conditions. It should be recalled that that the z-axis is along the interstellar axis.

It is seen from figures 8.8–8.10 that the conic-helical orbit of a planetoid moving with a relativistic velocity in the binary-star–planet system is relatively stable. Thus, under the influence of the heavier star in the binary system, the plane of the planetoid orbit, being at the lighter star, could orient itself to be perpendicular to the interstellar axis. The trajectory of the planet becomes corkscrew shaped.

All of the above celestial systems are capable of supporting a stable corkscrew motion of the lightest object in the system for a long, but not infinite, time. This fact does not diminish the significance of this discovery because *nothing exists for an infinite time*. Our Milky Way Galaxy will eventually crash into the nearby Andromeda Galaxy, although it will take few billion years. There are supermassive black holes at the centers of both galaxies and these black holes will ultimately smash together and merge into one larger black hole.

Incidentally, in 2018 for the first time a team of astronomers, led by Michael Koss from Eureka Scientific, actually observed several pairs of galaxies in the final stages of merging together into single, larger galaxies. They also observed the black holes in the centers of these galaxies drawing closer together before merging into one giant black hole. The team described their findings in a research paper published in the journal *Nature* [5].

To distract ourselves from the unpleasant thought of the inevitable death of our Milky Way Galaxy, here is a relevant joke:

> Two astrophysicists are discussing their research in a bar one evening when a drunk, sitting and listening on the next seat, turns and says, in a very worried voice: 'What was that you just said?'
>
> 'We were discussing stellar evolution, and I said to my colleague here that the Sun would run out of nuclear fuel and turn into a red giant star in about 5 billion years, possibly melting the Earth.'

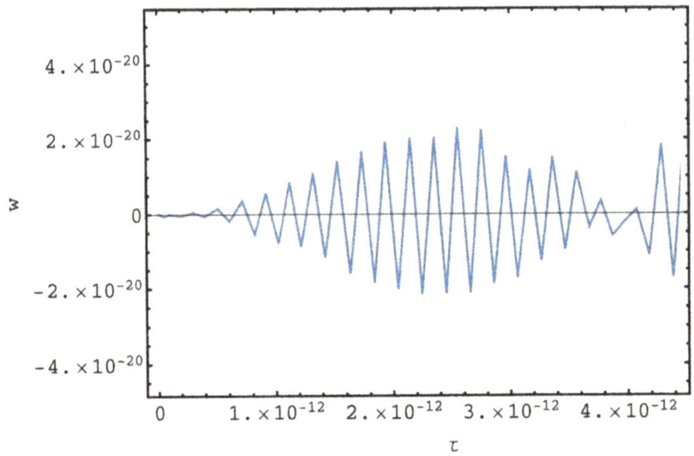

Figure 8.10. The same as in figure 8.8, but for the scaled z-coordinate of the planetoid $w = z/R$.

'Whew!' says the drunk, 'You really had me worried. I thought you said 5 million years.'

The title of this chapter revealed that these celestial systems have an analogy in the micro-world: one-electron Rydberg quasimolecules. So, let us discuss them in some detail.

One-electron Rydberg quasimolecules, consisting of two fully stripped ions of nuclear charges Z and Z' plus one highly excited electron, are encountered in various plasmas containing more than one kind of ion. Examples are (but are not limited to) magnetic fusion plasmas, laser-produced plasmas, plasmas used for lasers in the ultraviolet and x-ray ranges, solar plasmas and so on. In these plasmas, a fully stripped ion of nuclear charge Z' can come close to a hydrogen-like ion of nuclear charge Z and form a short-lived molecule (that is, a quasimolecule). Conversely, a fully stripped ion of nuclear charge Z can come close to a hydrogen-like ion of the nuclear charge Z' and form a quasimolecule. Such quasimolecules are a very useful playground for theoretical and experimental studies of charge exchange, which is a physical process of primary importance for many areas of physics—for example, for the areas listed above. Out of these research areas, the most practically important are studies of magnetic- and laser-fusion plasmas aimed at achieving an almost unlimited source of energy.

Classical analytical studies of one-electron Rydberg quasimolecules were presented in two papers published by the author of this book, in *Physical Review Letters* [6] and in *Journal of Physics B: Atomic, Molecular and Optical Physics* [7]. The primary result was the discovery of classical stable electronic orbits of the corkscrew shape. Namely, the electron trajectory is a helix on the surface of a cone, with the axis coinciding with the internuclear axis. In this *helical* state, the electron, while spiraling on the surface of the cone, oscillates between two end-circles which result from cutting the cone by two parallel planes perpendicular to its axis, see figure 8.11.

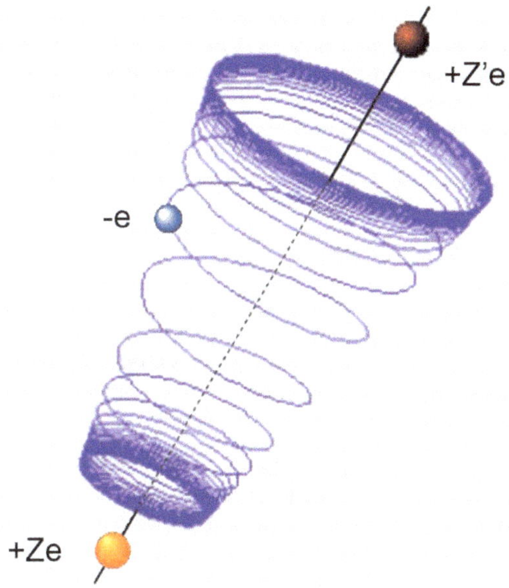

Figure 8.11. A sketch of the helical motion of the electron in the ZeZ'-system in the absence of a magnetic field. The trajectory is artificially stretched along the internuclear axis to make its details more visible. (Reproduced with permission from [8]. Copyright 2013 N Kryukov and E Oks.)

To be precise, it should be mentioned that the similarity between the above celestial systems and one-electron Rydberg quasimolecules is limited because of the following distinctions between these physical systems. First, in one-electron Rydberg quasimolecules the attractive centers (nuclei Z_1 and Z_2) can be stationary and still be stable, while in binary star systems the rotation of the stars is necessary for stability and in planet–star systems the rotation of the planet is necessary for stability.

Second, in one-electron Rydberg quasimolecules the attractive centers can engage in oscillations (called vibrations) and be stable without any rotation. This is not the case for the above celestial systems.

Despite these differences, it is still amazing how much the above celestial systems have in common with one-electron Rydberg quasimolecules. This analogy is quite counterintuitive.

Now is the time to explain why these micro-world systems are called quasimolecules rather than simply molecules. For this purpose let us recall the discussion from chapter 2 about using the shape of the potential energy for analyzing the stability (or lack thereof) of the relative motion of two objects. If the effective potential energy $U_{\text{eff}}(R)$ has a minimum at some distance R, such as in figure 2.2, then in the course of time, the distance r between the two objects could be confined between two turning points (corresponding in figure 2.2 to the crossings of the dashed horizontal line with the potential energy curve). However, if the effective potential energy $U_{\text{eff}}(R)$ does not have a minimum, then in the course of time the distance r between the two objects could become larger and larger—the two objects move away from each other forever.

So, let us apply this concept to one-electron Rydberg quasimolecules, considering, for example, a system of one electron around two nuclei of charges $Z = 2$ and $Z' = 3$. Since it is convenient to deal with scaled, dimensionless quantities, we will discuss here the shape of the dependence of the scaled, dimensionless effective potential energy

$$v = U_{\text{eff}} M^2 / Z^2, \qquad (8.2)$$

on the scaled, dimensionless internuclear distance

$$r = (Z/M^2)R, \qquad (8.3)$$

where M is the conserved projection of the electron angular momentum on the internuclear axis. Below we present the results published by Flannery and Oks in *Physical Review* A [9].

Figure 8.12 shows two branches of the dependence of the scaled effective potential energy v of the nuclear motion on the scaled internuclear distance r for $Z = 2$ and $Z' = 3$. It is seen that for any starting point on the lower branch, the system would 'find' a way to lower its potential energy without any obstacle and would end up at an infinitely large internuclear distance. The two nuclei are not bound together. This is why it is not a real molecule, but only a quasimolecule.

As for the upper branch in figure 8.12, the dependence $v(r)$ has a minimum, but it is very shallow. It is known from quantum mechanics that such shallow potential wells cannot confine the two nuclei. (This can be explained using the uncertainty relations discussed in chapter 3.) In addition, since such systems are usually embedded in a plasma, then due to the effects of the plasma environment, the shallow minimum could completely disappear. The net result is that again the two nuclei are not bound together, so that it is not a real molecule, but only a quasimolecule.

However, Flannery and Oks showed in [9] that a sufficiently strong magnetic field applied to a quasimolecule can bind the nuclei together and make a real molecule. This is called *magnetic stabilization of Rydberg quasimolecules.*

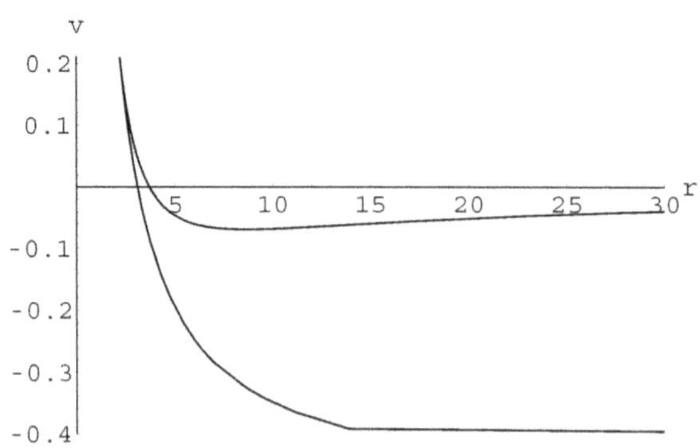

Figure 8.12. Two branches of the dependence of the scaled effective potential energy v of the nuclear motion (defined by equation (8.2)) on the scaled internuclear distance r (defined by equation (8.3)) for $Z = 2$ and $Z' = 3$.

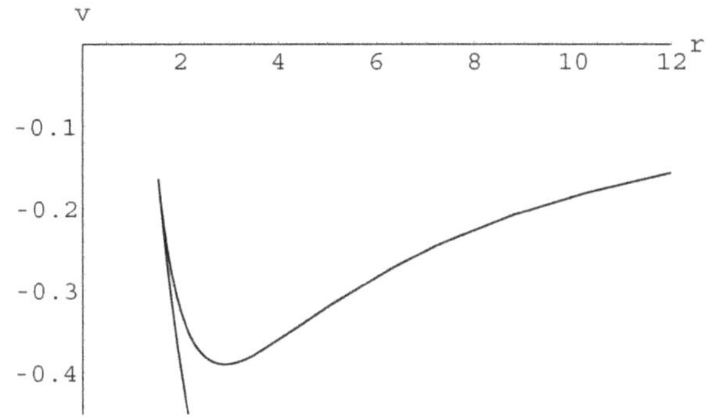

Figure 8.13. The same as in figure 8.12, but for the case of a relatively strong magnetic field applied along the internuclear axis.

Figure 8.13 shows the same as figure 8.12, but for the case of a relatively strong magnetic field applied along the internuclear axis. It is seen that the minimum in the upper branch became very pronounced. The magnetic field stabilizes the system and transforms the quasimolecule into a real molecule.

The particular example of the system whose scaled effective potential energy is presented in figures 8.12 and 8.13 corresponds to an electron in the vicinity of the nuclei of He and Li. Both nuclei are usually present in magnetic fusion plasmas. Moreover, in such plasmas, Rydberg states of either of these nuclei are always present in magnetic fusion plasmas (resulting from the so-called charge exchange with ions of higher nuclear charge). Relatively large magnetic fields are also present. It should therefore be possible to observe the magnetic stabilization of the quasimolecule $HeLi^{4+}$ present in these practically important experimental devices.

Flannery and Oks [9] showed that a similar magnetic stabilization of Rydberg quasimolecules is displayed by other (though not all) ZeZ'-systems. Namely, magnetic stabilization is possible for one-electron Rydberg quasimolecules characterized by a ratio of nuclear charges in the range $1 < Z'/Z < 3$.

References

[1] Beletsky V V 2001 *Essays on the Motion of Celestial Bodies* (Basel: Birkhäuser/Springer)
[2] Oks E 2015 *Astrophys. J.* **804** 106
Oks E 2016 *Astrophys. J.* **823** 69 (erratum)
[3] Kryukov N and Oks E 2015 *Int. Rev. At. Mol. Phys.* **6** 7
[4] Kryukov N and Oks E 2017 *J. Astrophys. Aerosp. Technol.* **5** 144
[5] Koss M J *et al* 2018 *Nature* **563** 214
[6] Oks E 2000 *Phys. Rev. Lett.* **85** 2084
[7] Oks E 2000 *J. Phys. B: At. Mol. Opt. Phys.* **33** 3319
[8] Kryukov N and Oks E 2013 *Int. Rev. At. Mol. Phys.* **4** 121
[9] Flannery R and Oks E 2006 *Phys. Rev.* A **73** 013405

IOP Publishing

Unexpected Similarities of the Universe with Atomic and Molecular Systems: What a Beautiful World

Eugene Oks

Chapter 9

The stormy relationship between black holes and atomic/quantum physics

Einstein's general relativity theory (that is, the theory of gravitation) predicted that a very massive (but relatively compact) object can curve space–time to such extent that neither particles nor electromagnetic radiation (light) can escape from it. The boundary of such a region is called the *event horizon*. Actually only in the 1960s was it shown that black holes are a generic prediction of general relativity.

Black holes are thought to originate from the collapse of very massive stars at the end of their lifetime. Then the black hole starts absorbing nearby stars and merging with other black holes. This leads to *supermassive black holes* of masses millions of times greater than the solar mass. It is commonly accepted that supermassive black holes exist at the centers of most galaxies, including our Milky Way Galaxy.

There is an alternative (albeit humorous) view on how black holes were created:

Black holes were created when God divided by zero!

While the interior of black holes is invisible, their location and even their masses can be determined by analyzing the orbit of a star (or stars) revolving around the black hole. Also their location and masses can be determined by the external matter falling into the black hole, causing the brightest objects in the Universe. As an unknown humorist wrote:

Black holes really suck!

Moreover, according to quantum mechanics (or more rigorously, to the quantum field theory in curved space–time), black holes emit some peculiar radiation called

doi:10.1088/2053-2563/aafed9ch9

Hawking radiation. It is named after the physicist Stephen Hawking who predicted it in 1974.

In a nutshell, the physics behind Hawking radiation is as follows. As we discussed in chapter 4, a vacuum is not actually empty: it is filled with 'zero-point' oscillations (that is, variations in time) of electric and magnetic fields. Close to the event horizon of a black hole, vacuum fluctuations give birth to a *particle–antiparticle* pair. (An antiparticle is a subatomic particle of the same mass as a given subatomic particle, but with an opposite electric charge or opposite magnetic properties.) One of them goes back into the black hole, but the other one escapes. As a result, the black hole gradually loses its mass and shrinks, and will eventually disappear, although this is an extremely slow process.

So, the physics of black holes combines relativity, gravity, and quantum mechanics. Incidentally, there are non-traditional definitions of some of these and other concepts, as follows:

Relativity: Family get-togethers at Christmas.
Gravity: The strength of a glass of beer.
Time travel: Throwing the alarm clock at the wall.
Black holes: What you get in black socks.

The lifetime of a particular black hole is much more probably limited by a collision and merger with another black hole, rather than by Hawking radiation. A merger of two black holes causes enormous ripples in space–time, propagating as gravitational waves. The first direct detection of such gravitational waves (and thus, of the merger of two black holes) occurred in 2016. In the following two years, five more mergers of black holes were observed by detecting the gravitational waves produced by the mergers. An unknown author wrote the following limerick:

A black hole—a tremendous creation,
Its physics defies imagination,
Time and space it can bend.
Wow! I can't comprehend
The gravity of this situation.

Here is another limerick on the same 'matter' (no pun intended):

There was an old man who observed,
'I confess, I am somewhat unnerved,
I had never before
Seen the truth of the lore
That, where matter is, space must be curved!

The life cycle of a black hole family is as follows. They merge with one another and create a supermassive black hole at the center of each galaxy. They gradually 'eat up' (that is, absorb) the stars and planets of each galaxy, so that all that remains of each galaxy is a supermassive black hole. Finally the supermassive black hole very slowly dies through the process of Hawking radiation.

But what would happen if somebody were to fall into a black hole. The answer below is totally counter-intuitive and mind-boggling. Below I follow an illuminating article by Amanda Gefner published on the BBC website in 2015 [1].

Let us assume that you have a space companion who remains safely on a spaceship outside the black hole. Here is what your companion would observe as you fall toward the black hole.

First of all, as you accelerate toward the event horizon, the companion sees you stretch and contort. Second, the closer you get to the event horizon, the more it appears to your companion that your motion slows down.

Third, but most importantly, when you reach the event horizon, your companion sees you stopped motionless, stretched across the surface of the horizon. He sees you slowly obliterated by the stretching of space, the stopping of time, and the fires of Hawking radiation. Eventually *he sees you burned to ashes.*

This apparent outcomes seems to be similar to what is described in the following limerick by A P French:

There was a young fellow named Cole,
Who ventured too near a black hole.
His dv by dt
Was quite wondrous to see,
But now all that's left is his soul.

(For clarity: 'dv by dt' means the time derivative of velocity, which is the acceleration.)

So, what do you think would be happening to you *from your point of observation*, rather than from your companion's point of observation? Would it be something just as horrible? Actually, something even stranger would happen—*nothing*!

You would not feel any stretching, slowing, bumping, or burning radiation. This is because you are in *free fall*. An object in free fall *does not feel any gravity*!

Free fall is when an object moves only with the force of gravity. Anybody who ever jumped up and down on a floor was in free fall for a split second. Anybody who jumped into a pool was in free fall for a fraction of a second. Anybody who jumped on a trampoline was in free fall for about a second. What you experienced in any of these situations was a very pleasant feeling of *weightlessness*.

Indeed, the sensation of weight in a gravitational field (such as, for example, the gravitational field of the Earth) is due to contact with supporting floors, seats, or scales and so on—rather than from the gravitational field itself. When the contact is removed (say, by jumping) and only the gravitational force remains, you experience weightlessness—like astronauts on the space station moving only under the gravitational force of the Earth.

The pleasant (some would say, happy) feeling of weightlessness does not depend on the strength of the gravitational force, but only on the absence of contact forces.

Therefore, inside a black hole you would only experience this pleasant feeling of weightlessness and nothing more.

The event horizon is not really a roadblock. Your companion outside the black hole cannot see through it, but (amazingly) from your point of view, there is *no horizon*!

You might have a problem if the black hole is relatively small. The force of gravity would be much greater at your feet than at your head. So, you would be disfigured, stretched like a wire.

However, in a supermassive black hole, this would not be the case. In a huge supermassive black hole, you could live quite normally, experiencing weightlessness, before dying at the very center of the black hole. At the very center of a black hole (called the singularity), the concept of space–time becomes meaningless and existence becomes impossible. But during the very long flight toward the center you would feel great.

So, how to reconcile this with the fact that your companion outside the black hole observes you burning to ashes? It was not a hallucination. In principle, *your companion could even collect your ashes* (!) and bring them back to Earth.

So, on the one hand, physics requires that, from the observation point of your companion, you must remain outside the black hole—because quantum physics necessitates that information can never be lost. The information about your condition has to remain outside the event horizon.

On the other hand, physics also requires that you fly through the event horizon without encountering anything. Otherwise, this would violate Einstein's general relativity.

Thus, the laws of physics require that you both be outside the black hole in the form of ashes and inside the black hole without any harm done to you. Moreover, physics requires that the information cannot be cloned. It might seem that you are simultaneously in two places. However, you can only see one copy of you. You and your companion cannot communicate. Nobody can see both inside and outside a black hole simultaneously. Are you really dead or are you really alive?

The answer is: *reality depends on the observation point.* There is your companion's reality and there is your reality—they coexist in two different regions of space–time.

Yet things are even more twisted. In principle, your companion might obtain some information about you from the inside the black hole—based on the phenomenon of *quantum entanglement.*

The concept of quantum entanglement relates to a pair of particles, such that the quantum state of each particle cannot be described independently of the state of the other, even when the particles are separated by a large distance. Instead, a quantum state must be described for the system as a whole. The physical properties of the two particles are *correlated.* Therefore, a measurement of some physical property of one particle (such as, for example, its angular momentum) automatically necessitates a particular value for the same property of the other one—regardless of how far away the other one is. Here is a humorous illustration.

When you pull one sock on your right foot, the second sock automatically becomes the left one. Moreover, it becomes the left sock instantaneously, regardless of the distance between your feet. This is the essence of the quantum entanglement.

However, your companion would need some kind of decoding machine to use quantum entanglement for connecting his bit of information outside the black hole to the bit of information inside the black hole, the latter bit being entangled with the first bit. Estimates show that even with the fastest computer allowed by the laws of physics (which does not yet exist), decoding would take your companion an astronomically huge amount of time, significantly exceeding the lifetime of the black hole.

So, your companion would not learn your side of the story. Unless he were also to choose to jump into the black hole.

Reference

[1] Gefner A 2015 The strange fate of a person falling into a black hole *BBC Earth* www.bbc.com/earth/story/20150525-a-black-hole-would-clone-you

Chapter 10

Atomic nuclei beyond the Standard Model and a natural explanation of the dark matter in the Universe

Atomic nuclei consist of two types of so-called nucleons: protons and neutrons. The nucleons are not 'elementary' particles—they are considered to consist of three elementary particles called *quarks*.

Quarks are the basis of the so-called Standard Model describing all the currently known 'elementary' particles. Quarks are somewhat mysterious because they have never been directly observed or found in isolation. A commonly accepted explanation is that as two quarks attempt to separate, at some extremely small distance it becomes energetically favorable to produce a new quark–antiquark pair rather than further continue the separation. (The antiquark is an antiparticle corresponding to a particular quark—like any antiparticle it has the same mass as its counterpart, but differs in the sign of some other property, such as, for example, the electric charge.) Information about quarks has been drawn indirectly, from observations of collisions of heavy 'elementary' particles. Of course, all quarks belong to the realm of quantum mechanics—specifically, to the part called quantum chromodynamics.

In 2001 Russian physicist Michael Romanovsky proposed an alternative, purely *classical* model of nucleons [1]. The motivation for developing his model was the absence of answers to several fundamental questions about nucleon properties, as follows. First, why do nucleons have the same value of so-called *spin* (which is an intrinsic angular momentum) as electrons or positrons (the latter being the electron antiparticle)? Second, why do protons have a 'strange' value of the magnetic moment, which is a non-integer in terms of the 'natural' unit of the magnetic moment (called the nuclear magneton) in nuclear physics? (Particles having a magnetic moment are like small bar magnets.) Third, why do neutrons (which are

doi:10.1088/2053-2563/aafed9ch10

electrically neutral) have a magnetic moment? Fourth, why is the ratio of the magnetic moments of neutrons and protons equal to 0.68? Fifth, why does a free neutron decay specifically in 881.5 s, yielding a proton, an electron, and an extremely light (much lighter than electrons) neutral particle called an antineutrino?

In Romanovsky's model, a nucleon consists of 252 small hard spheres filling about 0.7 of the nucleon volume. These small hard spheres (of yet unknown nature) move chaotically inside the nucleon and obey the laws of *classical* statistical physics.

The difference between protons and neutrons in his model is the following. In addition to 252 small hard spheres, inside the proton there is also a positron (the antiparticle of the electron). The positron is confined within the proton by yet unknown forces. As for the neutron, in addition to 252 small hard spheres and the confined positron, there is also an electron confined by the Coulomb attraction of the positron (that is, confined by the electric force).

Within this model, the mean value of the angular momentum of the chaotically moving 252 small hard spheres coincides with the experimental value of the intrinsic angular momentum (spin) of nucleons if the mean thermal velocity of the small hard spheres is 5.37×10^9 cm s^{-1}. This velocity is about six times smaller than the speed of light, which justifies the employment of the non-relativistic classical theory used by Romanovsky.

As for the magnetic moment according to his model, in the proton it arises due to the chaotic motion of the confined positron. In the neutron, the magnetic moment is due to the chaotic motion of the confined positron–electron pair.

For the proton, Romanovsky's model yields the experimentally observed value of the magnetic moment if the temperature of the chaotic motion of the confined positron is 6.05×10^4 eV. Since 1 eV is equal to 11 600 K (on the Kelvin temperature scale), then that temperature corresponds to 7.02×10^8 K.

For the neutron, where according to his model there is a confined positron–electron pair, a similar calculation yields the ratio of the magnetic moments of the neutron and the proton equal to 24/35, which is approximately 0.686. This practically coincides with the corresponding experimental ratio; it differs from the experimental one only in the third significant digit.

The finite lifetime of a free neutron in Romanovsky's model is due to the fact that the neutron can lose an electron by collision of the electron with the small hard spheres inside the neutron. His calculated lifetime of the neutron coincides with the experimental lifetime of 881.5 s if the small hard spheres fill about 0.7 of the neutron volume.

For a proton it is much more difficult to lose a positron by collisions with small hard spheres. Romanovsky's calculations yield a proton lifetime in the range of 10^{30}–10^{32} s. He pointed out that this range is in satisfactory agreement with estimates of the proton lifetime by other authors—about 10^{31} s.

Thus, Romanovsky's simple *classical* model based on some minimum number of assumptions yields the experimentally observed results for nucleon intrinsic angular momentum (spin) and the nucleon magnetic moments. The model also yields realistic estimates for the lifetime of neutrons and protons.

The small hard spheres within the nucleons represent unknown matter—unknown matter in the micro-world. In the Universe (that is, in the macro-world) there is also an unknown matter—called 'dark matter'—constituting about 85% of all the matter in the Universe. What if these two unknown matters are actually one and the same?

Actually, the author of this book has a pretty good idea what dark matter is. Hypotheses about dark matter, which were brought up before year 2018, either introduced new, undiscovered subatomic particles or proposed to dramatically change the known physical laws (for example, to modify Newtonian dynamics). The current author in his paper [2], published in 2018, suggested a more natural explanation for dark matter. It proceeds from the results of his previous paper [3], where he discovered theoretically and provided experimental evidence that hydrogen atoms could exist in two different 'flavors': the first flavor is the usual hydrogen atoms, but the second flavor is unusual hydrogen atoms. Namely, the second flavor of hydrogen atoms cannot emit or absorb electromagnetic radiation ('light') except for the radio line at wavelength 21 cm. Therefore, it is invisible for all kinds of astrophysical observations with just one exception, as follows.

Astrophysicists observe the redshifted radio line (which, without the redshift to larger wavelength, would have been at wavelength 21 cm) from the early Universe. They observed the absorption profile of this line: namely, as hydrogen atoms absorb radiofrequency photons from the cosmic microwave background. The second flavor of hydrogen atoms can significantly enhance such absorption.

In fact, this is probably what has been observed by Bowman *et al.* [4] in their paper published in 2018 before my paper [2]. They observed a significantly stronger absorption than could be explained by the usual kind (the first flavor) of hydrogen atoms. What they presented as a puzzling result, has been explained in paper [2]: the additional absorption was due to the second flavor of hydrogen atoms.

So, a nice thing about the second flavor of hydrogen atoms is that their existence *can be further tested by astrophysical observations.* Namely, further observational studies of the redshifted 21 cm radio line from the early Universe could provide further proof that dark matter or a part of it is the second flavor of hydrogen atoms.

Moreover, the existence of the second flavor of hydrogen atoms *already has a proof in laboratory atomic experiments.* Indeed, their existence helped explain in paper [3] a long-standing mystery of a very significant discrepancy between the experimental results on the high-energy tail of the linear momentum distribution in the ground state of hydrogen atoms and the previous theories. Further laboratory atomic experiments of this kind can provide further proof.

Thus, the existence of this kind of dark (or nearly dark) matter can be further verified both by astrophysical observations and by laboratory atomic experiments. This is a striking distinction to the previous hypotheses about dark matter—the hypotheses that had practically no way to be verified by either astrophysical observations or by laboratory experiments. Further details on this more natural, verifiable explanation of dark matter are presented in appendix E.

Let me finish this chapter with an appropriate joke by Thomas Hewitt Key:

What is mind?—No matter.
What is matter?—Never mind.

References

[1] Romanovsky M 2001 Preprint of the Institute of General Physics of the Russian Academy of Sciences, number 6

[2] Oks E 2018 *Int. Rev. At. Mol. Phys.* **9** 25

[3] Oks E 2001 *J. Phys. B: At. Mol. Opt. Phys.* **34** 2235

[4] Bowman J D, Rogers A E E, Monsalve R A, Mozdzen T J and Mahesh N 2018 *Nature* **555** 67

Chapter 11

Concluding remarks on the harmony and beauty of micro- and macro-worlds: symmetries ruling everything

The understanding and enjoyment of the harmony of our world has a long history. It may have started with the Greek philosopher and mathematician Pythagoras who lived in the 6th century BCE, that is, about 27 centuries ago. He introduced the concept of the 'harmony of the spheres'. He and his disciples believed that the distances between the planets in the solar system somehow corresponded to musical intervals. Twenty-one centuries later this concept inspired German mathematician and astronomer Johannes Kepler, who discovered and formulated the three laws of the motion of a planet around the Sun.

As for beauty, in physics and astronomy it is paired with simplicity. In search of the laws of nature, theoretical physicists can recognized truth by its beauty and simplicity, as Richard Feynman wrote in his book *The Character of Physical Law* [1], quoted in chapter 1.

The governing idea for theoretical physicists is to try to explain everything in nature from just a few principles. Most often, such principles are conservation laws —such as the conservation of energy, momentum, and angular momentum for a 'closed system'. A closed system is a system that does not interact with anything else in the Universe. Obviously, rigorously speaking, closed systems do not exist. However, if the internal interactions within a system are much stronger that its interactions with the rest of the world, then the closed system approach is a useful approximation, a useful *model*.

Models are the 'flesh and blood' of theoretical physics. When trying to explain new experimental facts, it is legitimate to develop a model that explains the new experimental data, as well as some (but not necessarily all) of the previously known

doi:10.1088/2053-2563/aafed9ch11

experimental findings. Models are the stepping stones to developing more comprehensive theories later on.

Moreover, there can be several different models to explain the same phenomenon. The plurality of models is beneficial. This is because if new experimental facts were to be found, one of these models might be better suited to embracing new experimental facts than other models.

Coming back to the conservation laws, they are actually consequences of *symmetries* in nature. In everyday language, a symmetry means a similarity (or sometimes exact correspondence) between different things. For example, the image in figure 11.1 (which to some might look like 'cat's ears') is symmetric with respect to the reflection about the vertical axis. This is because the two different parts of this image—the left part and the right part—look the same. Physicists would say that this image is *invariant* with respect to the reflection, that is, it does not change under reflection.

Reflection is just one of the possible *transformations* that physicists use to test symmetries. For example, if all particles in a closed system were to be moved in the same direction by the same distance (this is called 'parallel displacement'), the physics of the system could not change—because all locations in space are physically equivalent for a closed system. In other words, a closed system is invariant under a parallel displacement. The conservation of momentum for a closed system turns out to be a simple consequence of this invariance.

Similarly, for a closed system or for a system in a spherically symmetric field, all directions in space are equivalent. Therefore, the physics of the system cannot change under rotation of the entire system by some angle. The conservation of the angular momentum turns out to be a simple consequence of the invariance of such a system under rotation.

What about the conservation of energy for a closed system or for a system in a time-independent field? This turns out to be a simple consequence of the fact that all instants of time are equivalent for such a system (the invariance with respect to 'translation in time').

So, the conservation laws and the underlying symmetries are the cornerstone of physics. They provide deep insight into the fundamentals of nature.

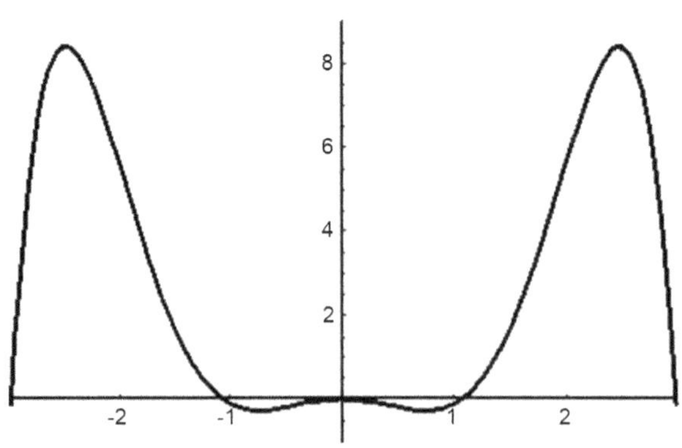

Figure 11.1. Example of a symmetric image.

In addition to the exact conservation laws and the underlying exact symmetries, there are also *approximate* conservation laws and the underlying *partial* symmetries. They are just as profound and useful as the exact ones. One example was given in chapter 7—a Rydberg atom in a high-frequency laser field. In this case, the exact symmetry is only axial (the axis being the direction of the laser field), so that the projection of the angular momentum on the direction of the laser field is exactly conserved. But since there is no spherical symmetry, the two other projections of the angular momentum (and therefore the square of the angular momentum vector) are not exactly conserved. However, it turns out that there is an approximate conservation of the square of the angular momentum in this system. This partial symmetry simplifies our understanding of the behavior of the system and the mathematical description of the behavior. The same partial symmetry was relevant to the celestial analogy of that atomic system—namely, to the motion of a satellite around the oblate Earth.

Another example of partial symmetry is the following. The so-called 'strong force'—one of the four fundamental forces in nature—holds together protons and neutrons inside nuclei. It turns out that the strong force between two protons is the same as between two neutrons. Moreover, it is the same as between a proton and a neutron. So, if one limits oneself to considerations of the strong force only, then protons and neutrons can be interchanged without observed consequences—thus, a partial symmetry. However, the symmetry is only partial because by also including the electric force the symmetry would be broken—protons differ from neutrons by the presence of the electric charge in protons.

The above examples illustrate the power of the inventiveness of theoretical physicists. As the joke goes:

A physicist can cross a wristwatch with helium so he can watch time fly.

In chapter 1, I promised to provide in the final chapter some comments on the so-called *computational physics* that should not be confused with theoretical physics. Typically, computational physicists, while trying to fit the experimental facts, develop computer programs (codes) based on fully numerical simulations starting from 'scratch' rather than from some analytical advance. These are called *simulations*.

There is no question about the importance of simulations as the third powerful research methodology—in addition to theories and experiments. Large-scale codes have been created to simulate a wide variety of complicated phenomena.

However, first, not all large-scale codes are properly verified and validated. Second, fully numerical simulations are generally ill-suited for capturing so-called emergent principles and phenomena, such as conservation laws, the laws of thermodynamics, detailed balance, and preservation of symmetries. Third, as with any fully numerical method, they lack physical insight.

A number of physicists started warning us about this several years ago. Let us present some relevant quotations.

In 2005 Post and Votta published a very insightful article in *Physics Today* [2], the main point of which was that 'much of computational science is still troublingly immature' and that new methods of verifying and validating complex codes are necessary and should be mandatory. Further they wrote:

> A computational simulation is only a model of physical reality. Such models may not accurately reflect the phenomena of interest. By verification we mean the determination that the code solves the chosen model correctly. Validation, on the other hand, is the determination that the model itself captures the essential physical phenomena with adequate fidelity. Without adequate verification and validation, computational results are not credible.

They described the underlying problems as follows:

> Part of the problem is simply that it's hard to decide whether a code result is right or wrong. Our experience as referees and editors tells us that the peer review process in computational science generally doesn't provide as effective a filter as it does for experiment or theory. Many things that a referee cannot detect could be wrong with a computational-science paper. The code could have hidden defects, it might be applying algorithms improperly, or its spatial or temporal resolution might be inappropriately coarse.

This was written by two leading experts in computational science. Indeed, Post is a computational physicist at Los Alamos National Laboratory and an associate Editor-in-Chief of the journal *Computing in Science and Engineering*. Votta is a Distinguished Engineer at Sun Microsystems and an Associate Editor of *IEEE Transactions on Software Engineering*. It is no wonder that their paper caused a lot of comments to be published in the following, August 2005, issue of *Physics Today*. In one of the comments, J Loncaric from Los Alamos wrote, in particular [3],

> Unfortunately, these days universities turn out users who employ codes as black boxes but do not understand what they do or when their results can be trusted.

Responding to the comments, Post and Votta, in the same issue of *Physics Today*, wrote [4]:

> Successful physical models capture the emergent principles that determine the behavior of complex systems. Examples of these emergent principles are conservation laws, the laws of thermodynamics, detailed balance, and preservation of symmetries… Nature includes all of the emergent phenomena, but a computational model may not.

The above general deficiencies of complicated codes resulted in huge failures of important large-scale projects. Post and Votta described the following examples.

> Examples abound of large-scale software failures in fields like information technology and aerospace. The 1995 failure of the European Space Organization's Arianne 5 rocket and the 1999 loss of NASA's Mars Climate Orbiter are still fresh in the memory. After the Columbia space shuttle's ill-fated February 2003 launch and the first reports of possible problems with the mission, a NASA—Boeing team's computational assessment of potential failure modes yielded misleading conclusions that may have contributed to the tragedy.
>
> The quest for fusion energy provides two more examples of problematic computation. By stretching boundary conditions far beyond what could be scientifically justified, computer simulations were able to 'reproduce' the exciting but wrong experimental discovery of sonoluminescent fusion. With regard to the International Thermonuclear Experimental Reactor (ITER), preliminary computational predictions in 1996 of inadequate performance by the proposed facility were wrongly characterized as definitive. Those predictions contributed to the 1998 US withdrawal from that important and promising international undertaking.

I have witnessed failures of some codes in other areas of atomic physics, such as, plasma spectroscopy. But I will not overload the readers with the specifics of these examples.

In summary, the fully numerical simulations of computational physics are often not properly verified and validated, *fail to capture conservation laws and symmetries*, and lack physical insight. Physics is an experimental science. So, the ultimate test of any theory should be comparison with experiments conducted in well-controlled conditions, rather than a comparison with unreliable simulations.

Finally let me emphasize the primary task of theoretical physics. While *quantitative* physical results are important, physics is first of all about the *qualitative* explanation of natural phenomena. For example, why does the Sun look red at sunrise and sunset? Why is the sky is blue? Why are grass and tree leaves green? Why do tree leaves turn yellow autumn? The answers are as follows.

'White' sunlight is a combinations of multiple colors, usually listed as red, orange, yellow, green, blue, indigo, and violet. Each color is an electromagnetic wave characterized by a particular frequency of the oscillations of the electric and magnetic fields in it. The frequency of blue light is about twice the frequency of red light (or in other words, the wavelength of blue light is twice as small as the wavelength of red light—since the wavelength and the frequency are inversely proportional to each other).

The air consists of molecules (mostly molecules of nitrogen and oxygen). When light propagates through air, it scatters on these molecules. The intensity of the scattering is proportional to the fourth power of the frequency. Therefore, the blue light scatters about 16 times more intensely than the red light ($2^4 = 16$).

So, when at sunrise or sunset we look directly at the Sun, the transmitted light is mostly red because all other colors (like blue) are scattered in all directions other than the direction of the observation.

When we look at the sky away from the position of the Sun, our eyes do not receive the light directly from the Sun, but only the light that is scattered on the molecules of the air. Therefore, the scattered light is mostly blue—because the red, orange, yellow, and green parts of sunlight do not scatter as intensely as the blue part. This is why the sky looks blue.

Let me quote a relevant limerick, by David Morin, Eric Zaslow, E'beth Haley, and Nathan Salwen:

A young child looked up in the sky,
And said, 'it's so blue, Mom, but why?'
You see, blue scatters more
(There's this power of four),
So it rarely comes straight to your eye.

Here is another limerick on the same topic, by Chris Mathews:

The sky has a beautiful hue,
To scattering of light it is due.
Long wavelengths like red,
Through air go unbled,
Leaving us seeing the blue!

There is even a whole poem (although less scientific) on this matter, by John Ciardi:

I don't suppose you happen to know
Why the sky is blue? It's because the snow
Takes out the white. That leaves it clean
For the trees and grass to take out the green.
Then pears and bananas start to mellow,
And bit by bit they take out the yellow.
The sunsets, of course, take out the red
And pour it into the ocean bed
Or behind the mountains in the west.
You take all this out and the rest
Couldn't be anything else but blue.
Look for yourself. You can see it's true.

Different colors are present in sunlight with different intensities. The color present in sunlight with the highest intensity is green. In order to absorb the maximum possible amount of solar energy, the grass is green and so are tree leaves in spring and summer.

In autumn the sky is cloudy much more frequently than in spring and summer. Each color of sunlight is partially absorbed by clouds. However, green light is absorbed by clouds more strongly than yellow light. Due to this fact, during autumn, the color reaching the Earth's surface with the highest intensity is yellow. In order to absorb the maximum possible amount of solar energy, tree leaves become yellow in autumn. (I never encountered this explanation of yellow leaves in autumns in the literature—so, I guess the explanation is mine.)

Apart from explaining natural phenomena, physics can save lives—according to the following joke:

> One day a professor was discussing a particularly complicated concept. A pre-med student rudely interrupted to ask: 'Why do we have to learn this pointless information?'
> 'To save lives', the professor responded quickly and continued the lecture.
> A few minutes later, the same student spoke up again: 'So, how does physics save lives?'
> 'It keeps the ignoramuses like you out of medical school', replied the professor.

Coming back to the harmony, beauty, and symmetries in nature, let me quote the French mathematician, physicist, and philosopher of science Jules-Henri Poincaré:

> The scientist does not study nature because it is useful; he studies it because he delights in it, and he delights in it because it is beautiful. If nature were not beautiful, it would not be worth knowing, and if nature were not worth knowing, life would not be worth living.

In my view, the most striking beauty in nature is the astonishing symmetry/similarities of the Universe with atomic and molecular systems, as presented in this book. I hope the reader starts sharing with me the feeling of the beauty of the world around us.

References

[1] Feynman R 2017 *The Character of Physical Law* (Cambridge, MA: MIT Press)
[2] Post D and Votta L 2005 *Phys. Today* **58** 35
[3] Loncaric J 2005 *Phys. Today* **58** 12
[4] Post D and Votta L 2005 *Phys. Today* **58** 12

Appendix A

Overview of generalized classical dynamics

Dirac [1–3] considered a dynamical system of N degrees of freedom characterized by generalized coordinates q_n and velocities $v_n = \mathrm{d}q_n/\mathrm{d}t$, where $n = 1, 2, \ldots, N$. If the Lagrangian of the system is

$$L = L(q, v), \tag{A.1}$$

then momenta are defined as

$$p_n = \partial L/\partial v_n. \tag{A.2}$$

Each of the quantities q_n, v_n, p_n can be varied by δq_n, δv_n, δp_n, respectively. The latter small quantities are of the order of ε, the variation being worked to the accuracy of ε. As a result of the variation, equation (A.2) would not be satisfied any more, since the right-hand side would differ from the corresponding left-hand side by a quantity of the order of ε.

Further, Dirac distinguished between two types of equations. To one type belong equations such as equation (A.2), which do not hold after variation (he called them 'weak' equations). Below for weak equations, following Dirac, we use an equality sign $\cong$ different from the usual sign. Another type constitutes equations such as equation (A.1), which holds exactly even after variation (Dirac called them 'strong' equations).

If quantities $\partial L/\partial v_n$ are *not* independent functions of velocities, one can exclude velocities v_n from equation (A.2) and obtain one or several weak equations

$$\phi(q, p) \cong 0, \tag{A.3}$$

containing only q and p. In his formalism, Dirac used the following complete system of independent equations of the type (A.3):

doi:10.1088/2053-2563/aafed9ch12

$$\phi_m(q, p) \cong 0, \quad (m = 1, 2, \ldots, M). \tag{A.4}$$

Here the word 'independent' means that neither of the ϕ can be expressed as a linear combination of the other ϕ with the coefficient depending on q and p. The word 'complete' means that any function of q and p, which would become zero allowing for equation (A.2) and which would change by ε under variation, should be a linear combination of the functions $\phi_m(q, p)$ from equation (A.4) with coefficients depending on q and p.

Finally, proceeding from the Lagrangian to a Hamiltonian, Dirac obtained the following central result:

$$H_g = H(q, p) + u_m\phi_m(q, p) \tag{A.5}$$

(here and below, summation over a twice repeated suffix is assumed). Equation (A.5) is a strong equation expressing a relation between the generalized Hamiltonian H_g and the conventional Hamiltonian $H(q, p)$. Quantities u_m are coefficients to be determined. Generally, they are functions of q, v, and p; by using equation (A.2), they could be made functions of q and p. It should be emphasized that $H_g \cong H(q, p)$ would only be a weak equation—in contrast to equation (A.5).

Equation (A.5) shows that the Hamiltonian is not uniquely determined, because a linear combination of ϕ may be added to it. Equations (A.4) are called *constraints*. The above distinction between constraints (i.e. weak equations) and strong equations can be reformulated as follows.

Constraints must be employed in accordance with certain rules. Constraints can be added. Constraints can be multiplied by factors (depending on q and p), but only on the left-hand side, so these factors must not be used inside Poisson brackets.

If f is some function of q and p, then df/dt (i.e. a general equation of motion) in the Dirac's generalized Hamiltonian dynamics (GHD) is

$$df/dt \cong [f, H] + u_m[f, \phi_m], \tag{A.6}$$

where $[f, g]$ is the usual Poisson bracket. Substituting $\phi_{m'}$ in equation (A.6) instead of f and taking into account equations (A.4), one obtains

$$[\phi_{m'}, H] + u_m[\phi_{m'}, \phi_m] \cong 0. \quad (m' = 1, 2, \ldots, M). \tag{A.7}$$

These consistency conditions allow determining the coefficients u_m.

We note that GHD was designed by Dirac specifically for applications to quantum field theory. In [4] Oks and Uzer showed how to apply GHD to atomic and molecular systems.

It is well-known that many dynamical systems possess *integrals of motion* other than energy. However, in our view, their role in atomic and molecular physics is not yet fully appreciated. Below we suggest using integrals of motion as the constraints in GHD.

We consider a classical atomic or molecular system of N degrees of freedom, possessing M classical integrals of motion $A_m(q, p)$, $m = 1, 2, \ldots, M$. We write the generalized Hamiltonian in the form (see equations (A.4) and (A.5))

$$H_g = H(q, p) + u_m\{A_m(q, p) - A_{0m}\}, \qquad A_{0m} = \text{const.} \tag{A.8}$$

Here A_{0m} is the value of $A_m(q, p)$ in a particular state of motion, so that in this state

$$A_m(q, p) - A_{0m} \cong 0. \tag{A.9}$$

Since the quantities $A_m(q, p)$ are integrals of motion, their Poisson bracket with $H(q, p)$ vanishes and the consistency condition (A.7) reduces to the form

$$u_m[A_{m'}, A_m] \cong 0. \qquad (m' = 1, 2, \ldots, M). \tag{A.10}$$

Equation (A.10) allows determining the coefficients u_m.

For a hydrogenic atom of nuclear charge Z, the integrals of the motion (other than energy) are the angular momentum $\mathbf{L} = \mathbf{r}\,\mathbf{p}$ and the Runge–Lenz vector [A.5] $\mathbf{A}(\mathbf{r}, \mathbf{p}) = \{\mathbf{r}p^2 - \mathbf{p}(\mathbf{r}\,\mathbf{p})\}/(\mu Z e^2) - \mathbf{r}/r$, where μ is the reduced mass. Therefore, the generalized Hamiltonian can be presented in the form

$$H_g = p^2/(2\mu) - Ze^2/r + \mathbf{u}\,(\mathbf{r}\,\mathbf{p} - \mathbf{L}_0) + \mathbf{w}\,(A(\mathbf{r}, \mathbf{p}) - \mathbf{A}_0), \tag{A.11}$$

where $\mathbf{L}_0$, $\mathbf{A}_0$, and the energy H_0 are connected by the well-known relation [5]

$$L_0^2 = \mu Z^2 e^4 (A_0^2 - 1)/(2H_0). \tag{A.12}$$

The consistency conditions $[\mathbf{r}\,\mathbf{p}, H_g] \cong 0$, $[\mathbf{A}(\mathbf{r}, \mathbf{p}), H_g] \cong 0$ allow one to express the Hamiltonian and the equations of motion in terms of just one unknown function $B(H_0, A_0)$:

$$\begin{aligned} H_g = p^2/(2\mu) - Ze^2/r + 2B(H_0, A_0)H_0\{\mathbf{M}_0\,(\mathbf{r}\,\mathbf{p})/M_0^2 \\ - (1 - \mathbf{A}_0\,\mathbf{A}(\mathbf{r}, \mathbf{p}))/(1 - A_0^2)\}, \end{aligned} \tag{A.13}$$

$$\begin{aligned} d\mathbf{r}/dt &= \{1 + B(H_0, A_0)\}\mathbf{p}/\mu, \\ d\mathbf{p}/dt &= -\{1 + B(H_0, A_0)\}Ze^2\mathbf{r}/r^3. \end{aligned} \tag{A.14}$$

The equations of motion (A.14) differ from their conventional form only by the factor $\{1 + B(H_0, A_0)\}$. Therefore, if we were to make a transformation of time

$$t' = \{1 + B(H_0, A_0)\}t, \tag{A.15}$$

then with respect to the new time t', the equations of the motion would be formally brought back to their conventional form.

Thus we come to the following *central point*. In the above generalized formalism, the trajectory of the atomic electron remains the same as in the conventional formalism. However, the generalized period T_g and the generalized frequency ω_g differ from their conventional values T_0 and ω_0 as follows:

$$T_g = T_0/|1 + B(H_0, A_0)|, \tag{A.16}$$

$$\omega_g = \omega_0|1 + B(H_0, A_0)| = |1 + B(H_0, A_0)||2H_0|^{3/2}/D^{1/2}, \qquad D \cong \mu Z^2 e^4. \tag{A.17}$$

It is seen from equation (A.17), that the generalized formalism allows the existence of such a state (or states) of motion, where $\omega_g = 0$ despite $H_0 \neq 0$ (the conventional formalism allows $\omega_0 = |2H_0|^{3/2}/D^{1/2} = 0$ only for $H_0 = 0$). This is a state where $B(H_0, A_0) = -1$. Therefore, such states would not emit electromagnetic radiation, would not lose energy to radiation, and would thus constitute stable states of the classical atom.

It is important to point out that the physics behind such classical non-radiating states is a *new kind of time-dilation* expressed by equation (A.16). This is a *non-Einsteinian time-dilation.* The closer the energy of the system to the energy of the classical non-radiating state, the more the time dilates. In the classical non-radiating state, time becomes infinitely dilated, so that the frequency ω_g in equation (A.17) vanishes and so does the radiation.

The function $B(H_0, A_0)$ can be determined based on the following experimental fact. Experiments show that a weakly bounded hydrogenic atom ($H_0 \to 0$) emits radiation primarily at some frequency tending to *a finite non-zero limit* as $H_0 \to 0$. (We use only the existence of such a limiting frequency of the radiation without using the experimental value for this limiting frequency.) Therefore, as $H_0 \to 0$, we should have from equation (A.17):

$$\omega_g \approx |B(H_0, A_0)||2H_0|^{3/2}/D^{1/2} = \text{const} \equiv Q, \tag{A.18}$$

such that

$$\mathrm{B}(H_0, A_0) = QD^{1/2}/|2H_0|^{3/2}. \tag{A.19}$$

Now we consider the energy $H_0 = H_{st}$ such that at this (yet unknown) energy there is no radiation because the frequency ω_g becomes zero. In accordance to the above, this occurs at

$$B(H_{st}, A_{st}) = -1. \tag{A.20}$$

Combining equations (A.19) and (A.20), we find $Q = -|2H_{st}|^{3/3}/D^{1/2}$, so that

$$B(H_0, A_0) = -|H_{st}/H_0|^{3/2}. \tag{A.21}$$

Thus, in the classical process of radiation from a weakly bounded state, the frequency of the radiation is

$$\omega_g(H_0) = (8/D)^{1/2}[|H_{st}|^{3/2} - |H_0|^{3/2}]. \tag{A.22}$$

Until the radiation vanishes, its frequency varies from $\omega_g^{\text{init}} \approx (8/D)^{1/2}|H_{st}|^{3/2}$ to $\omega_g^{\text{fin}} = 0$. Therefore, the *average* frequency in the process of radiation is

$$\langle\omega_g\rangle \approx (\omega_g^{\text{init}} + \omega_g^{\text{fin}})/2 \approx (2/D)^{1/2}|H_{st}|^{3/2}. \tag{A.23}$$

In accordance with Planck's hypothesis (later incorporated by Einstein into his photoelectric law—see, e.g. [6]), the change of the energy $\Delta E = |H_{st}| - |H_0| \approx |H_{st}|$ should be equal to $\hbar\langle\omega_g\rangle$:

$$|H_{st}| \approx \hbar\langle\omega_g\rangle \approx \hbar(2/D)^{1/2}|H_{st}|^{3/2}. \quad \text{(A.24)}$$

Thus we find the absolute value of the energy of the radiationless stable state to be

$$|H_{st}| = D/(2\hbar^2) = \mu Z^2 e^4/(2\hbar^2). \quad \text{(A.25)}$$

The value we obtain coincides exactly with the experimental result for the ionization potential of hydrogenic atoms. The function $B(H_0)$ is now completely defined by equations (A.21) and (A.25).

It should be emphasized that in the above derivation we employed two fundamental experimental facts, but did not 'forcefully' quantize any physical quantity describing the atom. In particular, this amounts to classically *deriving* the Bohr's postulate on the quantization of the angular momentum (instead of postulating it).

The above formalism allows one to deduce not only the lowest non-radiating stable state, but actually all other non-radiating stable states as well. To do this, we should recognize that the electron, while executing an elliptical orbit around the nucleus, emits classical electromagnetic radiation not only at the primary frequency of the revolution ω_g, but also at its harmonics (see, e.g. [7]): $\omega_r = k\omega_g$, $k = 1, 2, 3, \ldots$. Therefore, instead of equation (A.17), we should actually have

$$\omega_g = k|1 + B_k(H_0, A_0)||2H_0|^{3/2}/D^{1/2}. \quad \text{(A.26)}$$

As $H_0 \to 0$, we obtain from equations (A.26)

$$\omega_g \approx k|B_k(H_0, A_0)||2H_0|^{3/2}/D^{1/2} = \text{const} \equiv Q_k, \quad \text{(A.27)}$$

so that

$$B_k(H_0, A_0) = Q_k D^{1/2}/|2H_0|^{3/2}. \quad \text{(A.28)}$$

Now we consider the energy $H_0 = H_{st}(k)$ such that at this (yet unknown) energy there is no radiation because the frequency ω_r becomes zero. This occurs at

$$B_k(H_{st}(k), A_{st}(k)) = -1. \quad \text{(A.29)}$$

Combining equations (A.28) and (A.29), we find $Q_k = -|2H_{st}(k)|^{3/3}/D^{1/2}$, so that

$$B_k(H_0, A_0) = -|H_{st}(k)/H_0|^{3/2}. \quad \text{(A.30)}$$

In the classical process of radiation from a weakly bounded state, the frequency of the radiation is

$$\omega_g(H_0) = k(8/D)^{1/2}[|H_{st}|^{3/2} - |H_0|^{3/2}]. \quad \text{(A.31)}$$

Until the radiation vanishes, its frequency varies from $\omega_r^{init} \approx k(8/D)^{1/2}|H_{st}(k)|^{3/2}$ to $\omega_r^{fin} = 0$. Therefore, the *average* frequency in the process of the radiation is

$$<\omega_g> \approx (\omega_g^{init} + \omega_g^{fin})/2 \approx k(2/D)^{1/2}|H_{st}(k)|^{3/2}. \quad \text{(A.32)}$$

In accordance with Planck's hypothesis (later incorporated by Einstein into his photoelectric law—see, e.g. [6]), the change of the energy $\Delta E = |H_{st}(k)| - |H_0| \approx |H_{st}(k)|$ should be equal to $\hbar\langle\omega_g\rangle$:

$$|H_{st}(k)| \approx \hbar\langle\omega_r\rangle \approx \hbar k(2/D)^{1/2}|H_{st}(k)|^{3/2}. \tag{A.33}$$

Thus we find the absolute value of the energy of the non-radiating stable states to be

$$|H_{st}(k)| = D/(2\hbar^2k^2) = \mu Z^2e^4/(2\hbar^2k^2) \equiv -E_k, \quad k = 1, 2, 3, \ldots. \tag{A.34}$$

The value we obtained coincides exactly with the experimental result for the binding energy of stationary states of hydrogenic atoms. The integer k in equation (A.34) is not a 'mysterious' quantum number, but rather the number of the classical harmonic, at which the classical electromagnetic radiation is emitted by the atomic electron. The choice of

$$B_k(H_0) = -|E_k/H_0|^{3/2} \tag{A.35}$$

determines a Hamiltonian $H_g^{(k)}$ (see (A.13)), which classically reproduces a hydrogenic atom with a single non-radiating energy level k.

It is also feasible to classically reproduce a hydrogenic atom *with* N radiationless energy levels ($k = 1, 2, \ldots, N$) by choosing

$$B(H_0) = -\sum_{k=1}^{N}\theta(E_{k+1} - H_0)\,\theta(H_0 - E_k)|E_k/H_0|^{3/2}, \tag{A.36}$$

where $\theta(x)$ is the step-function: $\theta(x) = 1$ for $x > 0$, $\theta(x) = 0$ for $x \leqslant 0$.

Let us discuss a classical non-radiating stable state in more detail. In such a state we have $d\mathbf{r}/dt = d\mathbf{p}/dt = 0$, so that $\mathbf{r}(t) = \mathbf{r}_0$ and $\mathbf{p}(t) = \mathbf{p}_0$, where $\mathbf{r}_0$ and $\mathbf{p}_0$ are some vector constants. Thus, the electron is motionless, but its momentum *differs from zero*. This is not surprising—the momentum $\mathbf{p}$ is a more complex physical quantity than the velocity $\mathbf{v} \equiv d\mathbf{r}/dt$. For example, for a charge in an electromagnetic field characterized by a vector-potential $\mathbf{A}$, it is also possible to have $\mathbf{v} = [\mathbf{p} - e\mathbf{A}/(mc)]/m = 0$ while $\mathbf{p} = e\mathbf{A}/(mc) \neq 0$ [7].

It might seem that hydrogenic atoms would not constitute a 'typical' example of atomic and molecular physics because they belong to quantum systems, which have a higher symmetry than their geometrical symmetry and therefore possess 'additional' integrals of motion (sometimes called 'accidental' degeneracy). However, in reality, it turns out that, for instance, all *classical* spherically symmetric potentials have the O_4 and SU_3 symmetries [8–10]. Therefore, all such potentials possess an additional vector integral of motion (an analogue of the Runge–Lenz vector), which is an element of the Lie algebra and to which *no quantal operator can correspond* [8–10]. It is significant that the Poisson brackets of this classical vector integral of motion with other integrals of motion are the same as for the Runge–Lenz vector.

The most important is the following non-trivial fact. All classical systems of N degrees of freedom have the O_{N+1} and SU_N symmetries *regardless of the functional form of the Hamiltonian* [10]. In particular, all systems of three degrees of freedom (not only spherically symmetric ones) possess an additional vector integral of motion.

We emphasize again that for the overwhelming majority of microscopic systems, the additional ('non-geometrical') integrals of motion exist only classically, but not quantally (a few exceptions, such as the hydrogen atom, have been mentioned above). Thus, classical mechanics has an *intrinsic, built-in advantage* over quantum mechanics in this regard, but this advantage has never been employed to the fullest.

References

[1] Dirac P A M 1950 *Can. J. Math.* **2** 129
[2] Dirac P A M 1958 *Proc. R. Soc. London* **A246** 326
[3] Dirac P A M 1964 *Lectures on Quantum Mechanics* (New York: Academic) lecture 1
[4] Oks E and Uzer T 2002 *J. Phys. B: At. Mol. Opt. Phys.* **35** 165
[5] Landau L D and Lifshitz E M 1960 *Mechanics* (Oxford: Pergamon)
[6] Loeb L L 1938 *Atomic Structure* (London: Wiley) ch 6.2
[7] Landau L D and Lifshitz E M 1960 *Classical Theory of Fields* (Oxford: Pergamon)
[8] Bacry H, Ruegg H and Souriau J M 1966 *Commun. Math. Phys.* **3** 323
[9] Fradkin D M 1967 *Prog. Theor. Phys.* **37** 798
[10] Mukunda N 1967 *Phys. Rev.* **155** 1383

IOP Publishing

Unexpected Similarities of the Universe with Atomic and Molecular Systems: What a Beautiful World

Eugene Oks

Appendix B

Super-generalized Runge–Lenz vector

A so-called two-center problem (TCP) has the following two mathematically equivalent, but physically different, embodiments. The first is the motion of an electron in the field of two stationary Coulomb centers of charges Z and Z' separated by a distance R, which is one of the most fundamental problems in quantum mechanics (see, e.g. [1, 2]). The second is the motion of a planet in the gravitational field of two stationary stars of generally different masses, which is one of the most fundamental problems in celestial mechanics (see, e.g. [3, 4]). The geometrical symmetry of the TCP dictates the conservation of the energy and of the projection of the angular momentum on the internuclear (or interstellar) axis. It is also well-known that the TCP possesses a higher than geometrical symmetry and that there should be an additional integral of motion. The existence of the additional integral of motion is intimately connected with the fact that the TCP allows the separation of variables in the elliptical coordinates—this fact was shown as early as 1760 by Euler [5] (see also [6, 7]).

The Runge–Lenz vector $\mathbf{A}_0$ for one isolated Coulomb center, depicted in figure 3.4,

$$\mathbf{A}_0 = \mathbf{p} \times \mathbf{L} - Z\mathbf{r}/r \tag{B.1}$$

was known for centuries (atomic units are used throughout this appendix). Historically, it was introduced as early as in 1710 by Hermann [8, 9] and Bernoulli [10] and is therefore also called the Hermann–Bernoulli vector or Ermanno–Bernoulli vector (different just in the spelling of the first author's name, who is the same person in both cases)—more details on this history can be found in [11]. A century later it was rediscovered by Laplace, and then by Runge and Lenz (independently of each other).

doi:10.1088/2053-2563/aafed9ch13

Krivchenkov and Liberman [12] considered the TCP described by the Hamiltonian (or the Hamilton function)

$$H = p^2/2 - Z_1/r_1 - Z_2/r_2 + Z_2/R. \tag{B.2}$$

The charge Z_1 was placed at the origin and Z_2 at $z = R$. They presented the following expression for a super-generalized Runge–Lenz vector:

$$\mathbf{A}^{(KL)} = \mathbf{p} \times \mathbf{L} - Z_1\mathbf{r}_1/r_1 + Z_2\mathbf{r}_2/r_2 + Z_2\mathbf{e}_z, \tag{B.3}$$

where $\mathbf{e}_z$ is the unit vector in the direction from charge Z_1 to charge Z_2.

In [13] Kryukov and Oks calculated the Poisson bracket $[\mathbf{A}^{(KL)}, H]$ of the projection of this vector on the internuclear axis with the Hamiltonian from [12]. Surprisingly, the result was not zero:

$$[\mathbf{A}^{(KL)}, H] = -\left(2Z_1/r_1^3\right)\mathbf{r} \times \mathbf{L} \cdot \mathbf{e}_z - \left(2Z_2/r_2^3\right)R(xp_x + yp_y). \tag{B.4}$$

Thus, the projection of Krivchenkov–Liberman's vector on the internuclear axis is not conserved—contrary to their claim.

Gurarie [14] considered the TCP described by the Hamilton function:

$$H = p^2/2 - Z_1/r_1 - Z_2/r_2, \tag{B.5}$$

where Z_1 and Z_2 are placed into $z = a$ and $z = -a$, respectively (atomic units are also used here). He presented the following expression for a super-generalized Runge–Lenz vector:

$$\mathbf{A}^{(G)} = \mathbf{p} \times \mathbf{L} - Z_1(\mathbf{r} - a\mathbf{e}_z)/|\,\mathbf{r} - a\mathbf{e}_z\,| - Z_2(\mathbf{r} + a\mathbf{e}_z)/|\,\mathbf{r} + a\mathbf{e}_z\,|. \tag{B.6}$$

Kryukov and Oks [13] calculated the Poisson bracket $[\mathbf{A}^{(G)}, H]$ of the projection of this vector on the interstellar axis with the Hamiltonian from [14]. Surprisingly, again the result was not zero:

$$[\mathbf{A}^{(G)}, H] = -2a(xp_x + yp_y)\left(Z_1/|\,\mathbf{r} - a\mathbf{e}_z\,|^3 - Z_2/|\,\mathbf{r} + a\mathbf{e}_z\,|^3\right). \tag{B.7}$$

Thus, the projection of Gurarie's vector on the interstellar axis is not conserved—contrary to his claim.

To derive a correct expression for the super-generalized Runge–Lenz vector, Kryukov and Oks [13] started from the Hamiltonian/Hamilton function given by equation (B.2) and followed the first few steps from Krivchenkov–Liberman's paper [12], arriving at the same expression in the elliptical coordinates (v, w, ϕ), as they did, for the additional conserved quantity:

$$\begin{aligned} A = {} & 1/R1/(w^2 - v^2)\left((w^2 - 1)(1 - v^2)\left(p_w{}^2 - p_v{}^2\right) - ((w^2 - 1)/(1 - v^2)\right. \\ & \left. - (1 - v^2)/(w^2 - 1))p_\varphi{}^2\right) - Z(1 + wv)/(w + v) \\ & + Z'(1 + v)(w - 1)/(w - v). \end{aligned} \tag{B.8}$$

Then, using the relation between the elliptical coordinates and the Cartesian coordinates

$$x = (R/2)\,((w^2 - 1)(1 - v^2))^{1/2}\cos\varphi,$$
$$y = (R/2)\,((w^2 - 1)(1 - v^2))^{1/2}\sin\varphi, \quad z = (R/2)wv, \tag{B.9}$$

Kryukov and Oks [13] obtained from equation (B.8) the super-generalized Runge–Lenz vector, whose projection on the internuclear axis is really conserved, and which has the form

$$\mathbf{A} = \mathbf{p}\times\mathbf{L} - L^2/R\mathbf{e}_z - Z\mathbf{r}/r - Z'(\mathbf{R} - \mathbf{r})/|\,\mathbf{R} - \mathbf{r}\,| + Z'\mathbf{e}_z, \quad \mathbf{e}_z = \mathbf{R}/R. \tag{B.10}$$

The Poisson bracket of the projection of the vector from equation (B.10) on the internuclear axis

$$A_z = \mathbf{p}\times\mathbf{L}\cdot\mathbf{e}_z - L^2/R - Zz/r - Z'(R - z)/|\,\mathbf{R} - \mathbf{r}\,| + Z' \tag{B.11}$$

with the Hamilton function from equation (B.2) indeed vanishes.

In the limit of large R, the expression from equation (B.11) reduces to the following asymptotic form

$$(A_z)_{\rm as} = \mathbf{p}\times\mathbf{L}\cdot\mathbf{e}_z - L^2/R - Zz/r + (1/2)(x^2 + y^2)Z'/R^2. \tag{B.12}$$

Here the direction of the z-axis is chosen from charge Z to charge Z'.

In 1964 Redmond [15] considered the problem of a hydrogen atom in a uniform electric field **F** for the purpose of deriving a generalized Runge–Lenz vector **A** for this simpler problem. His resultant vector **A** and its projection A_z on the axis $Oz||\mathbf{F}$ can be represented, respectively, in the forms

$$\mathbf{A} = \mathbf{A}_0 + (\mathbf{r}\times\mathbf{F})\times\mathbf{r}/2, \qquad A_z = A_{0z} - (x^2 + y^2)F/2. \tag{B.13}$$

Compared to Redmond's result from equation (B.13), there is an extra term in the Kryukov–Oks equation (B.12): $-L^2/R$. At first glance, this might seem puzzling. However, the Poisson bracket of $-L^2/R$ with the Hamilton function from equation (B.2) yields a term of the order of $1/R^3$. Thus, the Poisson bracket of $A_{\rm as}$ from equation (B.12) with the Hamilton function from equation (B.2), being calculated up to (including) terms of the order of $1/R^2$, vanishes. This resolves what might have seemed to be a puzzle.

The correct super-generalized Runge–Lenz vector for the TCP that we derived should be of general theoretical interest since the TCP is one of the most fundamental problems in physics. It can also have practical applications; for example, it can be used as a necessary tool while applying to the TCP the robust perturbation theory for degenerate states (based on the integrals of motion) developed by Oks and Uzer [16].

References

[1] Power J D 1973 *Philos. Trans. R. Soc. Lond.* A **274** 663

[2] Komarov I V, Ponomarev L I and Yu Slavyanov S 1976 *Spheroidal and Coulomb Spheroidal Functions* (Moscow: Nauka) (in Russian)

[3] Charlier C L 1927 *Die Mechanik des Himmels, (Celestial Mechanics)* (Berlin: de Gruyter)
[4] Pars L A 1965 *A Treatise on Analytical Dynamics* (London: Heinemann)
[5] Euler L 1760 *Mem. De Berlin* 228
[6] Born M 1925 *Vorlesungen über Atommechanik (The Mechanics of the Atom)* (Berlin: Springer)
[7] Landau L D and Lifshitz E M 1960 *Mechanics* (Oxford: Pergamon)
[8] Hermann J 1710 *Giornale de Letterati D'Italia* **2** 447
[9] Hermann J 1712 *Hist. Acad. R. Sci. Mem. Math. Phys.* **1712** 519
[10] Bernoulli J I 1712 *Hist. Acad. R. Sci. Mem. Math. Phys.* **1712** 685
[11] Goldstein H 1976 *Am. J. Phys.* **44** 1123
[12] Krivchenkov V D and Liberman M A 1968 *Russ. Phys. J.* **11** 14
[13] Kryukov N and Oks E 2012 *Phys. Rev.* A **85** 054503
[14] Gurarie D 1992 *Symmetries and Laplacians: Introduction to Harmonic Analysis, Group Representations and ApplicationsNorth Holland Mathematics Studies* vol 174 (Amsterdam: North Holland)
[15] Redmond P J 1964 *Phys. Rev.* B **133** 1352
[16] Oks E and Uzer T 2000 *Europhys. Lett.* **49** 554

Appendix C

The classical Stark effect

The classical Hamilton function (for brevity, Hamiltonian) of a charged particle of mass m and charge q moving in a Coulomb field and subjected to a *weak* uniform electric field **F** can be written in the form

$$H = p^2/(2m) - \alpha/r - q\mathbf{Fr}. \tag{C.1}$$

In this appendix, for any two vectors **A** and **B**, the notation **AB** stands for the scalar product (also known as the dot-product) of these two vectors. Below we follow in part the presentation from the Kotkin–Serbo book [1].

The time derivative of the angular momentum **M**, calculated through the Poisson bracket of vector **M** with the Hamiltonian (C.1), yields

$$\mathrm{d}\mathbf{M}/\mathrm{d}t = q\mathbf{r} \times \mathbf{F}, \tag{C.2}$$

where $\mathbf{r} \times \mathbf{F}$ is the vector product (also known as the cross-product) of these two vectors. The time derivative of the Runge–Lenz vector $\mathbf{A} = \mathbf{v} \times \mathbf{M} - \alpha\mathbf{r}/r$, calculated through the Poisson bracket of vector **A** with the Hamiltonian (C.1), yields

$$\mathrm{d}\mathbf{A}/\mathrm{d}t = (q/m)\mathbf{F} \times \mathbf{M} + q\mathbf{v} \times [\mathbf{r} \times \mathbf{F}]. \tag{C.3}$$

(It should be noted that the above definition of vector **A** from book [1] differs from the standard definition of the Runge–Lenz vector by the coefficient $-1/\alpha$, but this is not significant for the purposes of this appendix.)

In accordance to classical perturbation theory, equations (C.2) and (C.3) have to be averaged over the period of the unperturbed motion around the Kepler ellipse. As a result, one obtains

doi:10.1088/2053-2563/aafed9ch14

$$d\mathbf{M}_{av}/dt = [3q/(4|E|)]\mathbf{F} \times \mathbf{A}_{av}, \tag{C.4}$$

$$d\mathbf{A}_{av}/dt = [3q/(2m)]\mathbf{F} \times \mathbf{M}_{av}, \tag{C.5}$$

where E is the energy and subscript 'av' stands for 'averaged'.

Let us proceed first with a particular case where the electric field **F** is perpendicular to the angular momentum vector **M**, meaning that the vector **F** is located in the plane of the orbit. In this case the orbit remains in the plane, the vector **M** keeps its direction (apart from the sign), and the following equation can be obtained for the magnitude of the vector $\mathbf{M}_{av}$

$$d\mathbf{M}_{av}/dt = [3e\alpha q\mathbf{F}/(4|E|)]\sin\psi, \tag{C.6}$$

where

$$e = [1 - 2\,|\,E\,|\,\mathbf{M}^2/(m\alpha^2)]^{1/2} \tag{C.7}$$

is the eccentricity of the Kepler ellipse and ψ is the angle between the vectors **A** and **F**. Taking into account that

$$e\cos\psi = \varepsilon = \text{const} \tag{C.8}$$

and excluding e and ψ from equation (C.6), one obtains

$$\mathbf{M}_{av}(t) = M_0\cos(\Omega t + \beta), \tag{C.9}$$

where

$$\Omega = 3q\mathrm{F}/[2(2m|E|)^{1/2}], \tag{C.10}$$

$$M_0 = [m\alpha^2(1 - \varepsilon^2)/(2|E|)]^{1/2}. \tag{C.11}$$

The time evolution of the eccentricity is given by

$$e(t) = [1 - (1 - \varepsilon^2)\cos^2(\Omega t + \beta)]^{1/2}. \tag{C.12}$$

The corresponding trajectory is the ellipse swinging around the direction of the electric field **F** and changing the eccentricity in phase with its swinging, see figure 6.5. As the direction of the angular momentum **M** changes its sign (according to equation (C.9)), the direction of the motion of the particle around the ellipse changes to the opposite one.

The above solution is valid as long as the (formally obtained) expression for the swinging frequency Ω from equation (C.10) remains much smaller than the primary frequency of the motion, which is the Kepler frequency of the particle revolution around the ellipse. This imposes a limit on the strength of the electric field **F**—the field must remain relatively small.

In the general case of an arbitrary orientation of the electric field **F** with respect to the plane of orbit, one has to find the common solution of equations (C.4) and (C.5).

It is easily seen from equations (C.4) and (C.5) that the components of the vectors $\mathbf{M}_{av}$ and $\mathbf{A}_{av}$ parallel to the vector $\mathbf{F}$ are conserved:

$$\mathbf{F}\mathbf{M}_{av} = \text{const}, \quad \mathbf{F}\mathbf{A}_{av} = \text{const}. \tag{C.13}$$

The equation for the perpendicular component $\mathbf{M}_{av,per}$ of the vector $\mathbf{M}_{av}$, can be obtained in the form

$$d^2\mathbf{M}_{av,per}/dt^2 + \Omega^2\mathbf{M}_{av,per} = 0. \tag{C.14}$$

By introducing the Cartesian reference frame with the z-axis along the electric field $\mathbf{F}$, the obvious solution of equation (C.14) can be represented in the form:

$$\mathbf{M}_{av,x}(t) = B_1 \cos \Omega t + C_1 \sin \Omega t, \tag{C.15}$$

$$\mathbf{M}_{av,y}(t) = B_2 \cos \Omega t + C_2 \sin \Omega t, \tag{C.16}$$

where constants B_1, B_2, C_1, and C_2 are determined by the initial values of the vectors $\mathbf{M}_{av}$ and $\mathbf{A}_{av}$.

By substituting equations (C.15) and (C.16) in equation (C.5), one obtains a similar result for the perpendicular components of the vector $\mathbf{A}_{av}$:

$$\mathbf{A}_{av,x}(t) = -[3qF/(2m\Omega)](B_2 \sin \Omega t - C_2 \cos \Omega t), \tag{C.17}$$

$$\mathbf{A}_{av,x}(t) = [3qF/(2m\Omega)](B_1 \sin \Omega t - C_1 \cos \Omega t), \tag{C.18}$$

Reference

[1] Kotkin G L and Serbo V G 1971 *Collection of Problems in Classical Mechanics* (Oxford: Pergamon) problem 2.32

Appendix D

Derivation of the effective potential energy of a hydrogen Rydberg atom in a high-frequency laser field

The first part of this appendix contains a brief overview of the Nadezhdin–Oks paper [1]. The starting point is the same as in section 30 of the Landau–Lifshitz book [2]. Let us consider a particle in a static potential $U(x_\alpha)$ and under the force $\mathbf{f}(x_\alpha)\cos\omega t$. We seek the solution of the equation of motion $md^2(x_\alpha)/dt^2 = -dU/dx_\alpha + f_\alpha \cos\omega t$ in the form $x_\alpha(t) = X_\alpha(t) + \xi_\alpha(t)$ and expand the right-hand side in powers of the rapid oscillations ξ_α:

$$\begin{aligned} md^2(X_\alpha)/dt^2 + md^2(\xi_\alpha)/dt^2 = &- dU/dX_\alpha - \xi_\alpha d^2U/dX_\alpha dX_\beta \\ &- (1/2)\xi_\beta\xi_\gamma d^3U/dX_\alpha dX_\beta dX_\gamma \\ &+ \left[f_\alpha(\mathbf{X}) + \xi_\beta df_\alpha/dX_\beta\right]\cos\omega t. \end{aligned} \tag{D.1}$$

Here and below, the summation over repeated subscripts is assumed. In contrast to section 30 of [2], in equation (D.1) the term $-(1/2)\,\xi_\beta\xi_\gamma\, d^3U/dX_\alpha dX_\beta dX_\gamma$ is taken into account.

For the oscillatory terms it is sufficient to write $md^2(\xi_\alpha)/dt^2 = f_\alpha(\mathbf{X})\cos\omega t$, so that $\xi_\alpha = -[f_\alpha/(m\omega^2)]\cos\omega t$. Substituting the latter equation in equation (D.1) and averaging over the period $2\pi/\omega$, we obtain the equation for the averaged motion $\mathbf{X}(t)$:

$$\begin{aligned} md^2(X_\alpha)/dt^2 = &- dU/dX_\alpha - [1/(4m^2\omega^4)]f_\beta f_\gamma d^3U/dX_\alpha dX_\beta dX_\gamma \\ &- [1/(2m\omega^2)]f_\beta\, df_\alpha/dX_\beta. \end{aligned} \tag{D.2}$$

doi:10.1088/2053-2563/aafed9ch15

If the gradients of the force $f_\alpha(\mathbf{X})$ are relatively large, then in equation (D.2) it is possible to disregard the term proportional to $1/\omega^4$ compared to the term proportional to $1/\omega^2$, as was done in section 30 of [2]. However, for a spatially uniform force **f**, the term proportional to $1/\omega^2$ vanishes, so that it becomes necessary to take into account the term proportional to $1/\omega^4$. As a result, the equation for the averaged motion takes the form

$$m\mathrm{d}^2(X_\alpha)/\mathrm{d}t^2 = -\mathrm{d}/\mathrm{d}X_\alpha\left\{U + \left[f_\beta f_\gamma/(4m^2\omega^4)\right]\mathrm{d}^2U/\mathrm{d}X_\beta\mathrm{d}X_\gamma\right\}. \tag{D.3}$$

Thus, in this situation we deal with an effective potential

$$U_{\text{eff}} = U + \left[f_\beta f_\gamma/(4m^2\omega^4)\right]\mathrm{d}^2U/\mathrm{d}X_\beta\mathrm{d}X_\gamma. \tag{D.4}$$

By applying this result to a hydrogen Rydberg atom in a *linearly polarized* high-frequency laser field, Nadezhdin and Oks [1] obtained

$$U_{\text{eff}}(r,\theta) = -e^2/r - (\gamma_0/r^3)(3\cos^2\theta - 1), \qquad \gamma_0 = e^4E_0^{\,2}/(4m_e^{\,2}\omega^4), \tag{D.5}$$

where θ is the polar angle, that is, the angle between the radius-vector **r** of the electron and the z-axis chosen along the vector-amplitude $\mathbf{E}_0$ of the laser field. This equation is reproduced in chapter 7 as equation (7.6).

Oks, Davis, and Uzer [3] applied the effective potential from equation (D.4) to the general case of a hydrogen atom in an *elliptically polarized* high-frequency laser field:

$$\mathbf{E}(t) = \mathbf{e}_x\varepsilon_0\cos\,\omega t + \mathbf{e}_y\mu\varepsilon_0\sin\omega t, \tag{D.6}$$

where μ is the ellipticity degree. The peak field ε_0 in expression (D.6) is connected with the time-average of the electric field as follows:

$$<E^2(t)> = <\{\varepsilon_0^{\,2}\cos^2\omega t + \mu^2\varepsilon_0^{\,2}\sin^2\omega t\}> = \varepsilon_0^{\,2}(1+\mu^2)/2. \tag{D.7}$$

On the other hand, one can define an effective amplitude E_0 through

$$<E^2(t)> = E_0^{\,2}/2. \tag{D.8}$$

By equating the previous two expressions, one obtains the relation between E_0 and ε_0:

$$\varepsilon_0 = E_0/(1+\xi^2)^{1/2}. \tag{D.9}$$

By applying the effective potential from equation (D.4) to the case of the high-frequency laser field from equation (D.6), Oks, Davis, and Uzer obtained the following effective potential:

$$U_{\text{eff}} = -e^2/r + (\gamma/r^3)\{(1+\mu^2) - 3\sin^2\theta[1-(1-\mu^2)\sin^2\varphi]\}, \tag{D.10}$$

where

$$\gamma = (e^4\varepsilon_0^{\,2})/(4m^2\omega^4), \tag{D.11}$$

θ is the polar angle, and φ is the azimuthal angle of the radius-vector of the electron (the z-axis is chosen to be perpendicular to the polarization plane).

In the particular case where the high-frequency laser field is *circularly polarized* ($\mu = 1$), the effective potential from equation (D.7) simplifies to

$$U_{\text{eff}}(r, \theta) = -e^2/r + (\gamma/r^3)(3\cos^2\theta - 1). \quad \text{(D.12)}$$

Equation (D.12) with γ from equation (D.11) is reproduced in chapter 7 as equation (7.9).

For this case of circular polarization, by using equation (D.9) with $\mu = 1$, equation (D.11) for γ can also be expressed through the time-average square of the laser field as follows:

$$\gamma = (e^4 E_0^2)/(8m^2\omega^4). \quad \text{(D.13)}$$

References

[1] Nadezhdin B B and Oks E 1986 *Sov. Tech. Phys. Lett.* **12** 512

[2] Landau L D and Lifshitz E M 1976 *Mechanics* (Amsterdam: Elsevier)

[3] Oks E, Davis J E and Uzer T 2000 *J. Phys. B: At. Mol. Opt. Phys.* **33** 207

Appendix E

A natural explanation of dark matter

From astrophysical observations, it is clear that in the Universe there is over five times more of the unknown, dark matter than the known, ordinary matter. Existing hypotheses about dark matter either introduce new, undiscovered subatomic particles or propose to dramatically change the known physical laws (such as the modified Newtonian dynamics or related proposals)—see, e.g., reviews [1–3], and the book [4], and references therein. The current author in his recent paper [5] suggested a more natural explanation for dark matter. It proceeds from the results of his previous paper [6] dealing with singular solutions of the Dirac equation, as follows.

Solutions of the Dirac equation for the electron in the Coulomb field are common eigenfunctions of four operators (as is well-known—see, e.g., the textbook [7]—the Hamiltonian H, the projection J_z of the total angular momentum, the square J^2 of the total angular momentum, and the following operator:

$$K = \beta(2\mathbf{Ls} + 1). \tag{E.1}$$

Here β is the Dirac matrix of the rank four, whose nonzero elements are $\beta_{11} = \beta_{22} = 1$, $\beta_{33} = \beta_{44} = -1$; $\mathbf{L}$ and $\mathbf{s}$ are the operators of the orbital angular momentum and spin, respectively; $\mathbf{Ls}$ denotes the dot-product (also known as the scalar product) of the latter two operators. Eigenvalues of the operators K and J^2 are connected as follows: $k = \pm(j + 1/2)$.

Hydrogen atoms in the stationary states have the following well-known energies

$$E_{\mathrm{Nk}} = mc^2\{1 + \alpha^2/[N + (k^2 - \alpha^2)^{1/2}]^2\}^{-1/2}, \tag{E.2}$$

where $\alpha = e^2/(\hbar c)$ is the fine structure constant and N is the radial quantum number. For the ground state, the quantum numbers N and k have the following values

$$N = 0, \quad k = -1, \tag{E.3}$$

so that

$$E_{0,-1} = mc^2(1 - \alpha^2)^{1/2}. \tag{E.4}$$

For hydrogen atoms the radial part $R_{\mathrm{Nk}}(r)$ of the coordinate wave functions has the following behavior at small r (see, e.g., the textbook [7]):

$$R_{\mathrm{Nk}}(r) \propto 1/r^{1+s}, \qquad s = \pm(k^2 - \alpha^2)^{1/2}. \tag{E.5}$$

For the ground state of hydrogen atoms (hereafter, GSHA), equation (E.5) reduces to:

$$R_{0,-1}(r) \propto 1/r^{\mathrm{q}}, \qquad q = 1 \pm (1 - \alpha^2)^{1/2}. \tag{E.6}$$

In paper [6] it was shown, that with the allowance for the finite proton size, both the regular exterior solution corresponding to $q = 1 - (1 - \alpha^2)^{1/2}$ and the singular exterior solution corresponding to $q = 1 + (1 - \alpha^2)^{1/2}$ are legitimate for the GSHA[1].

The *central point* of paper [5] is the following. On the one hand, for hydrogen atoms both the regular ground state and the singular ground state correspond to the same energy given by equation (E.4) with the same quantum numbers $N = 0$, $k = -1$. In other words, the ground state of hydrogen atoms, in addition to the trivial double-degeneracy with respect to the z-projection m_{j} of the total angular momentum J, has an *additional double-degeneracy*.

On the other hand, there is a fundamental theorem of quantum mechanics concerning the cause of any additional degeneracy. Namely, it is caused by the existence of an additional conserved quantity (or quantities), whose operator commutes with the Hamiltonian, but does not commute with other conserved quantities or if it does, but the additional conserved quantity is a multi-component one, then its components do not commute with each other—see, e.g., the textbook [8]. The corresponding degenerate states differ by the eigenvalues (i.e., by additional quantum numbers) of the additional conserved quantity. For example, in the non-relativistic Coulomb problem, the higher than geometrical degeneracy (the geometrical one being due to the spherical symmetry of the problem) is caused by the existence of the Runge–Lenz vector **A**, whose operator commutes with the Hamiltonian, but does not commute with some components of the angular momentum operator (and the components of the Runge–Lenz vector operator do not commute with each other)—see, e.g., the textbook [8]. For a given energy, the corresponding non-relativistic degenerate states differ not only by the eigenvalue of the operator L_z, but also by the eigenvalue of the operator A_z.

[1] Here and below, by 'singular' we mean the strongly-singular solution of the Dirac equation for the Coulomb field—in distinction to the commonly accepted 'regular' solution that has a weak singularity at the origin.

However, according to paper [6], the GSHA is double-degenerate despite these two states having the same values of all the known conserved quantities E, K, J^2, and J_z. This means that *there should be an additional, new conserved quantity, by eigenvalues of which these two ground states differ*. In other words, the situation is that hydrogen atoms have *two flavors*, differing by the eigenvalue of an additional, new conserved quantity: hydrogen atoms have *flavor symmetry*.

Speaking of flavors: quarks (such as, for example, the constituents of protons and neutrons) have flavors—for instance, there are up and down quarks. For representing this particular flavor symmetry, there was assigned an operator of the isotopic spin (isospin) I—the operator having two eigenvalues for its z-projection: $I_z = 1/2$ assigned to the up quark and $I_z = -1/2$ assigned to the down quark.

By analogy with the isospin, in paper [5] there was introduced a new operator: the operator of *isohydrogen spin*, abbreviated as *isohyspin* and denoted as $I^{(h)}$. Similarly to the isospin, the z-projection of the isohyspin operator has two eigenvalues: $I^{(h)}{}_z = 1/2$ assigned to the regular flavor of hydrogen atoms and $I^{(h)}{}_z = -1/2$ assigned to the singular flavor of hydrogen atoms.

The isospin couples to the strong force (strong interaction). This is logical because it is related to intra-nuclear physics, where the strong interaction plays the dominant role. As a result, the strong force can transform the up quark into the down quark and vice versa.

In distinction, the isohyspin does not relate to intra-nuclear physics: so, it would be logical to state that the isohyspin does not couple to the strong force/interaction—since the isohyspin relates to a hydrogen atom as a whole. For the same reason, it would be logical to state that the isohyspin does not couple to the weak force/interaction. Also, there seems no grounds to expect that the isohyspin would couple to the gravitational force/interaction. As for the electromagnetic force/interaction, the (ordinary) spin couples to the magnetic field, but the isospin does not couple to the electromagnetic force/interaction. Therefore, there seems to be no reason for the isohyspin to couple to the electromagnetic force/interaction either.

It should be emphasized that in paper [6] there was presented also the first *experimental proof* of the existence of the second flavor of hydrogen atoms. Namely, there existed a long-standing mystery of the huge discrepancy between the experimental and previous theoretical results concerning the high-energy tail of the linear momentum distribution in the ground state of hydrogen atoms. The previous theories predicted the tail to scale with the linear momentum p as $\sim 1/p^6$, while the experiments pointed to the scaling of $\sim 1/p^k$ with the value of k close to 4. It was shown in paper [6] that the allowance for the second flavor of hydrogen atoms eliminates this huge discrepancy. Thus, there were already both the theoretical and the experimental proofs of the existence of the second flavor of hydrogen atoms.

So, in paper [5] it was suggested that *dark matter or a part of it could be represented by the second flavor of hydrogen atoms*. This is due to the fact that the second flavor of hydrogen atoms does not have states required for the emission or absorption of the electromagnetic radiation—because such states cannot be described by the singular solution outside the proton. (The only possible radiative transitions for the singular flavor of hydrogen atoms are between the hyperfine structure substates of the ground state, the latter transition being known as the radio line of 21 cm wavelength[2].) This idea seems to provide a more natural explanation for dark matter—because it does not require introducing new, undiscovered subatomic particles or dramatically changing the known physical laws.

Moreover, the existence of the second flavor of hydrogen atoms could explain a puzzling observational result published in 2018 in *Nature* by Bowman *et al* [9]. The authors of paper [9] observed the 21 cm line (redshifted from the rest frequency of 1240 MHz to the frequency of 78 MHz) from the early Universe. They observed the absorption profile of this line: namely, as hydrogen atoms absorb photons from the cosmic microwave background (CMB). The underlying physical mechanism was the ultraviolet light from stars formed in the early Universe—the light that is expected to penetrate the primordial hydrogen gas and to alter the excitation of the hydrogen 21 cm hyperfine structure line.

The puzzling result from paper [9] was that the amplitude of the profile was more than a factor of two greater than the largest predictions. This could mean that the primordial hydrogen gas was much cooler than expected.

The intensity of the observable 21 cm line from the early Universe is given as the brightness temperature T_B, which is a linear combination of the CMB temperature T_{CMB} and the spin temperature T_S (the latter being the excitation temperature of the hyperfine transition). In its turn, the spin temperature T_S is, generally speaking, a linear combination of the following three contributions (see, e.g., paper [10] and review [11]): the CMB temperature T_{CMB}, the gas kinetic temperature T_K (due to the fact that there is the collisional excitation of the hyperfine transition), and the color temperature of the radiation field in the Lyman series T_{Ly}. The latter contribution is due to the Wouthuysen–Field effect. Physically, the Wouthuysen–Field effect is the transition between the hyperfine structure sublevels of the ground state facilitated by the absorption and the subsequent re-emission of a photon of the Lyman series—mostly the Ly-alpha photon.

A possible explanation of the puzzling observational result from paper [9] was provided in paper [5] as follows. Hydrogen atoms of the second flavor do not have states required for the absorption or emission of Lyman quanta. Therefore, for such atoms the contribution, which would have been due to the Wouthuysen–Field effect, is absent. Therefore, the spin temperature of the hydrogen atoms of the second flavor is significantly smaller than for the hydrogen atoms in the regular flavor. The effective spin temperature of the mixture of the two flavors would be dominated by the spin temperature of the singular flavor since it could be expected that the latter flavor is the dominating part of the mixture—consistent with the fact that dark matter is considered to be more abundant than ordinary matter. The smaller effective spin temperature leads to the greater amplitude of the 21 cm absorption signal observed in [12]. This explanation seems to be more natural than resorting to a possible cooling of baryons by unspecified dark matter particles, as in papers [12, 13].

---

[2] For this reason, rigorously speaking, the singular flavor of hydrogen atoms could be called nearly-dark matter.

In summary, there is the second flavor of hydrogen atoms: they do not have states required for the emission or absorption of the electromagnetic radiation—except for the hyperfine structure substates of the ground state (the latter transition being known as the radio line of 21 cm wavelength). Therefore, *the second flavor of hydrogen atoms could represent dark matter (or nearly-dark matter) or a part of it.* In this way, dark matter or a part of it could be explained without resorting to new subatomic particles or without dramatically changing the physical laws.

As for the radiative transitions between the hyperfine structure substates of the ground state (corresponding to the radio line of 21 cm), the second flavor of hydrogen atoms could explain the puzzling observational results by Bowman *et al* [9]—without introducing an unspecified dark matter. Therefore, further observational studies of the redshifted 21 cm radio line from the early Universe could provide further proof that dark matter or a part of it is the singular flavor of hydrogen atoms.

In addition to the astrophysical observations, atomic experiments for accurately measuring the high-energy tail of the linear momentum distribution in the ground state of hydrogen atoms could also provide further proof of the existence of the singular flavor of hydrogen atoms. Thus, a nice thing about the second flavor of hydrogen atoms is that while the first experimental proof of their existence had been already given in paper [6], further proofs can be provided by astrophysical observations, as well as by laboratory experiments.

References

[1] Bertone G and Tait T M P 2018 *Nature* **562** 51
[2] Bertone G and Hooper D 2018 *Rev. Mod. Phys.* **90** 045002
[3] Garrett K and Düda G 2011 *Adv. Astron.* **2011** 968283
[4] Majumdar D 2014 *Dark Matter: An Introduction* (Boca Raton, FL: CRC Press)
[5] Oks E 2018 *Int. Rev. At. Mol. Phys.* **9** 25
[6] Oks E 2001 *J. Phys. B: At. Mol. Opt. Phys.* **34** 2235
[7] Rose M W 1961 *Relativistic Electron Theory* (New York: Wiley)
[8] Landau L D and Lifshitz E M 1965 *Quantum Mechanics* (Oxford: Pergamon)
[9] Bowman J D, Rogers A E E, Monsalve R A, Mozdzen T J and Mahesh N 2018 *Nature* **555** 67
[10] Zaldarriaga M, Furlanetto S R and Hernquist L 2004 *Astrophys. J.* **608** 622
[11] Furlanetto S R, Oh S P and Briggs F H 2006 *Phys. Rep.* **433** 181
[12] Barkana R 2018 *Nature* **555** 71
[13] Tashiro H, Kadota K and Silk J 2014 *Phys. Rev.* D **90** 083522

Lightning Source UK Ltd.
Milton Keynes UK
UKHW052313090920
369545UK00003B/91

9 780750 317054